ClearRevise™

AQA GCSE
Chemistry 8462 / 8464

Illustrated revision and practice

Foundation and Higher
Chemistry and Trilogy Courses

Published by
PG Online Limited
The Old Coach House
35 Main Road
Tolpuddle
Dorset
DT2 7EW
United Kingdom

sales@pgonline.co.uk
www.clearrevise.com
www.pgonline.co.uk
2021

PREFACE

Absolute clarity! That's the aim.

This is everything you need to ace the examined component in this course and beam with pride. Each topic is laid out in a beautifully illustrated format that is clear, approachable and as concise and simple as possible.

Each section of the separate Chemistry and combined science (Trilogy) specifications are clearly indicated to help you cross-reference your revision. The checklist on the contents pages will help you keep track of what you have already worked through and what's left before the big day.

We have included worked exam-style questions with answers for almost every topic. This helps you understand where marks are coming from and to see the theory at work for yourself in an exam situation. There is also a set of exam-style questions at the end of each section for you to practise writing answers for. You can check your answers against those given at the end of the book.

LEVELS OF LEARNING

Based on the degree to which you are able to truly understand a new topic, we recommend that you work in stages. Start by reading a short explanation of something, then try and recall what you've just read. This has limited effect if you stop there but it aids the next stage. Question everything. Write down your own summary and then complete and mark a related exam-style question. Cover up the answers if necessary but learn from them once you've seen them. Lastly, teach someone else. Explain the topic in a way that they can understand. Have a go at the different practice questions – they offer an insight into how and where marks are awarded.

ACKNOWLEDGEMENTS

The questions in the ClearRevise textbook are the sole responsibility of the authors and have neither been provided nor approved by the examination board.

Every effort has been made to trace and acknowledge ownership of copyright. The publishers will be happy to make any future amendments with copyright owners that it has not been possible to contact. The publisher would like to thank the following companies and individuals who granted permission for the use of their images in this textbook.

Design and artwork: Jessica Webb / PG Online Ltd
Photographic images: © Shutterstock, (p125) © sciencephotos/Alamy Stock Photo
Calorimeter heat test, © Thermtest inc., www.thermtest.com

First edition 2021
A catalogue entry for this book is available from the British Library
ISBN: 978-1-910523-32-2
Contributor: Nigel Saunders
Copyright © PG Online 2021
All rights reserved
No part of this publication may be reproduced, stored in a retrieval system, or transmitted in any form or by any means without the prior written permission of the copyright owner.

Printed on FSC certified paper by Bell and Bain Ltd, Glasgow, UK.

THE SCIENCE OF REVISION

Illustrations and words

Research has shown that revising with words and pictures doubles the quality of responses by students.[1] This is known as 'dual-coding' because it provides two ways of fetching the information from our brain. The improvement in responses is particularly apparent in students when asked to apply their knowledge to different problems. Recall, application and judgement are all specifically and carefully assessed in public examination questions.

Retrieval of information

Retrieval practice encourages students to come up with answers to questions.[2] The closer the question is to one you might see in a real examination, the better. Also, the closer the environment in which a student revises is to the 'examination environment', the better. Students who had a test 2–7 days away did 30% better using retrieval practice than students who simply read, or repeatedly reread material. Students who were expected to teach the content to someone else after their revision period did better still.[3] What was found to be most interesting in other studies is that students using retrieval methods and testing for revision were also more resilient to the introduction of stress.[4]

Ebbinghaus' forgetting curve and spaced learning

Ebbinghaus' 140-year-old study examined the rate in which we forget things over time. The findings still hold power. However, the act of forgetting things and relearning them is what cements things into the brain.[5] Spacing out revision is more effective than cramming – we know that, but students should also know that the space between revisiting material should vary depending on how far away the examination is. A cyclical approach is required. An examination 12 months away necessitates revisiting covered material about once a month. A test in 30 days should have topics revisited every 3 days – intervals of roughly a tenth of the time available.[6]

Summary

Students: the more tests and past questions you do, in an environment as close to examination conditions as possible, the better you are likely to perform on the day. If you prefer to listen to music while you revise, tunes without lyrics will be far less detrimental to your memory and retention. Silence is most effective.[5] If you choose to study with friends, choose carefully – effort is contagious.[7]

---

1. Mayer, R. E., & Anderson, R. B. (1991). Animations need narrations: An experimental test of dual-coding hypothesis. *Journal of Education Psychology*, (83)4, 484–490.
2. Roediger III, H. L., & Karpicke, J.D. (2006). Test-enhanced learning: Taking memory tests improves long-term retention. *Psychological Science*, 17(3), 249–255.
3. Nestojko, J., Bui, D., Kornell, N. & Bjork, E. (2014). Expecting to teach enhances learning and organisation of knowledge in free recall of text passages. *Memory and Cognition*, 42(7), 1038–1048.
4. Smith, A. M., Floerke, V. A., & Thomas, A. K. (2016) Retrieval practice protects memory against acute stress. *Science*, 354(6315), 1046–1048.
5. Perham, N., & Currie, H. (2014). Does listening to preferred music improve comprehension performance? *Applied Cognitive Psychology*, 28(2), 279–284.
6. Cepeda, N. J., Vul, E., Rohrer, D., Wixted, J. T. & Pashler, H. (2008). Spacing effects in learning a temporal ridgeline of optimal retention. *Psychological Science*, 19(11), 1095–1102.
7. Busch, B. & Watson, E. (2019), *The Science of Learning*, 1st ed. Routledge.

CONTENTS

Paper 1

Topic 1 — Atomic structure and the periodic table

Chemistry	Trilogy			☑
4.1.1.1	5.1.1.1	Atoms, elements and compounds	2	☐
4.1.1.1, 4.2.2.2	5.1.1.1, 5.2.2.2	Chemical equations	3	☐
4.1.1.1	5.1.1.1	Half equations and ionic equations	4	☐
4.1.1.2	5.1.1.2	Separating mixtures	5	☐
4.1.1.2	5.1.1.2	Crystallisation	6	☐
4.1.1.2	5.1.1.2	Chromatography	7	☐
4.1.1.2	5.1.1.2	Simple distillation	8	☐
4.1.1.2	5.1.1.2	Fractional distillation	9	☐
4.1.1.3	5.1.1.3	Developing the atomic model	10	☐
4.1.1.4–5	5.1.1.4–5	Subatomic particles	11	☐
4.1.1.5	5.1.1.5	Isotopes	12	☐
4.1.1.6	5.1.1.6	Relative atomic mass	13	☐
4.1.1.7	5.1.1.7	Electronic structure	14	☐
4.1.2.1	5.1.2.1	The periodic table	15	☐
4.1.2.2	5.1.2.2	The development of the periodic table	16	☐
4.1.2.3	5.1.2.3	Metals and non-metals	17	☐
4.1.2.4	5.1.2.4	Group 0	18	☐
4.1.2.5	5.1.2.5	Group 1	19	☐
4.1.2.6	5.1.2.6	Group 7	20	☐
4.1.2.6	5.1.2.6	Halogen displacement reactions	21	☐
4.1.3.1		Transition metals vs Group 1 | **Chemistry only**	22	☐
4.1.3.2		Charges, colours, and catalysts | **Chemistry only**	23	☐
		Examination practice 1	**24**	☐

Topic 2 — Bonding, structure, and the properties of matter

Chemistry	Trilogy			☑
4.2.1.1	5.2.1.1	Chemical bonds	26	☐
4.2.1.2	5.2.1.2	Ionic bonding	27	☐
4.2.1.3	5.2.1.3	Ionic compounds	28	☐
4.2.1.4	5.2.1.4	Covalent bonding	29	☐
4.2.1.4	5.2.1.4	Dot and cross diagrams	30	☐
4.2.1.5	5.2.1.5	Metallic bonding	31	☐
4.2.2.1	5.2.2.1	The three states of matter	32	☐
4.2.2.1	5.2.2.1	Using the particle model	33	☐
4.2.2.3	5.2.2.3	Properties of ionic compounds	34	☐

4.2.2.4–5	5.2.2.4–5	Properties of small molecules	35
4.2.2.5	5.2.2.5	Polymers	35
4.2.2.6	5.2.2.6	Giant covalent structures	36
4.2.2.7–8	5.2.2.7–8	Properties of metals and alloys	37
4.2.3.1–2	5.2.3.1–2	Diamond and graphite	38
4.2.3.3	5.2.3.3	Graphene and fullerenes	39
4.2.4.1		Nanoparticles | **Chemistry only**	40
4.2.4.2		Uses of nanoparticles | **Chemistry only**	41
		Examination practice 2	**42**

Topic 3   Quantitative chemistry

Chemistry ■   **Trilogy** ■

4.3.1.1–2	5.3.1.1–2	Relative formula mass	44
4.3.1.2	5.3.1.2	Percentage composition by mass	45
4.3.1.3	5.3.1.3	Mass changes with gases	46
4.3.1.4	5.3.1.4	Chemical measurements	47
4.3.2.1	5.3.2.1	Moles | **Higher Tier only**	48
4.3.2.2	5.3.2.2	Amounts of substances in equations | **Higher Tier only**	49
4.3.2.3	5.3.2.3	Balancing equations using moles | **Higher Tier only**	50
4.3.2.4	5.3.2.4	Limiting reactants | **Higher Tier only**	51
4.3.2.5	5.3.2.5	Concentration of solutions	52
4.3.3.1		Percentage yield | **Chemistry only**	53
4.3.3.2		Atom economy | **Chemistry only**	54
4.3.4		Concentrations in mol/dm^3 | **Chemistry (Higher only)**	55
4.3.5		Volumes of gases | **Chemistry (Higher only)**	56
		Examination practice 3	**57**

Topic 4   Chemical changes

Chemistry ■   **Trilogy** ■

4.4.1.1, 4.4.1.4	5.4.1.1, 5.4.1.4	Oxidation and reduction	59
4.4.1.2	5.4.1.2	The reactivity series	60
4.4.1.3	5.4.1.3	Extraction of metals	61
4.4.2.1	5.4.2.1	Reactions of acids with metals	62
4.4.2.2	5.4.2.2	Formulae of ionic compounds	63
4.4.2.2	5.4.2.2	Neutralisation of acids	64
4.4.2.3	5.4.3.2	Salts from insoluble reactants	65
8.2.1 RPA 1	10.2.8 RPA 8	Preparing a pure, dry soluble salt	66
4.4.2.4	5.4.2.4	The pH scale	67
4.4.2.4	5.4.2.4	Investigating neutralisation	68
4.4.2.5		Titrations | **Chemistry only**	69
8.2.2 RPA 2		Titration	70

v

4.4.2.5		Titration calculations | **Chemistry (Higher Tier only)**	71 ☐
4.4.2.6	5.4.2.5	Strong and weak acids | **Higher Tier only**	72 ☐
4.4.3.1, 4.4.3.5	5.4.3.1, 5.4.3.5	Electrolysis	73 ☐
4.4.3.2	5.4.3.2	Electrolysis of molten ionic compounds	74 ☐
4.4.3.3	5.4.3.3	Extracting metals using electrolysis	75 ☐
4.4.3.4–5	5.4.3.4–5	Electrolysis of aqueous solutions	76 ☐
8.2.3 RPA 3	**10.2.9 RPA 9**	**Investigating electrolysis**	77 ☐
		Examination practice 4	**78** ☐

Topic 5   Energy changes

Chemistry ■ **Trilogy** ■ ☑

4.5.1.1	5.5.1.1	Exothermic and endothermic reactions	80 ☐
8.2.4 RPA 4	**10.2.10 RPA 10**	**Exothermic and endothermic reactions**	81 ☐
4.5.1.2	5.5.1.2	Reaction profiles	82 ☐
4.5.1.3	5.5.1.3	The energy change of reactions | **Higher Tier only**	83 ☐
4.5.2.1–2		Chemical cells and fuel cells | **Chemistry only**	84 ☐
		Examination practice 5	**86** ☐

Paper 2

Topic 6   The rate and extent of chemical change

Chemistry ■ **Trilogy** ■ ☑

4.6.1.1	5.6.1.1	Calculating rates of reactions	88 ☐
8.2.5 RPA 5	**10.2.11 RPA 11**	**Measuring rate of reaction**	90 ☐
4.6.1.2	5.6.1.2	Factors affecting the rate of reaction	92 ☐
4.6.1.3	5.6.1.3	Collision theory	93 ☐
4.6.1.3–4	5.6.1.3–4	Activation energy and catalysts	94 ☐
4.6.2.1–2	5.6.2.1–2	Reversible reactions	96 ☐
4.6.2.3	5.6.2.3	Equilibrium	97 ☐
4.6.2.4	5.6.2.4	Le Chatelier's Principle | **Higher Tier only**	98 ☐
4.6.2.5	5.6.2.5	Concentration changes and equilibrium | **Higher Tier only**	99 ☐
4.6.2.6	5.6.2.6	Temperature changes and equilibrium | **Higher Tier only**	100 ☐
4.6.2.7	5.6.2.7	Pressure changes and equilibrium | **Higher Tier only**	101 ☐
		Examination practice 6	**102** ☐

Topic 7   Organic chemistry

Chemistry ■ **Trilogy** ■ ☑

4.7.1.1	5.7.1.1	Crude oil, hydrocarbons and alkanes	104 ☐
4.7.1.2	5.7.1.2	Fractional distillation and petrochemicals	105 ☐

Chemistry	Trilogy			
4.7.1.3	5.7.1.3	Properties of hydrocarbons	106	☐
4.7.1.4	5.7.1.4	Cracking	107	☐
4.7.1.4	5.7.1.4	Alkenes	108	☐
4.7.2.1		Structure and formulae of alkenes | **Chemistry only**	109	☐
4.7.2.2		Reactions of alkenes | **Chemistry only**	110	☐
4.7.2.2		Alkenes and water | **Chemistry only**	112	☐
4.7.2.3		Alcohols | **Chemistry only**	113	☐
4.7.2.3		Fermentation | **Chemistry only**	114	☐
4.7.2.4		Carboxylic acids | **Chemistry only**	115	☐
4.7.2.4		Reactions of carboxylic acids | **Chemistry only**	116	☐
4.7.3.1		Addition polymerisation | **Chemistry only**	117	☐
4.7.3.2		Condensation polymerisation | **Chemistry (Higher Tier only)**	118	☐
4.7.3.3–4		Amino acids and DNA | **Chemistry only**	119	☐
		Examination practice 7	**120**	☐

Topic 8   Chemical analysis

Chemistry	Trilogy			☑
4.8.1.1	5.8.1.1	Pure substances	122	☐
4.8.1.2	5.8.1.2	Formulations	123	☐
4.8.1.3	5.8.1.3	Explaining chromatography	124	☐
8.2.6 RPA 6	10.2.12 RPA 12	Paper chromatography	125	☐
4.8.2.1–4.8.2.4	4.8.2.1–4.8.2.4	Identifying common gases	126	☐
4.8.3.1		Flame tests | **Chemistry only**	127	☐
4.8.3.2		Metal hydroxides | **Chemistry only**	128	☐
4.8.3.3–4.8.3.5		Identifying anions | **Chemistry only**	129	☐
8.2.7 RPA 7		Chemical analysis	130	☐
4.8.3.6		Instrumental methods of analysis | **Chemistry only**	131	☐
4.8.3.7		Flame emission spectroscopy | **Chemistry only**	132	☐
		Examination practice 8	**133**	☐

Topic 9   Chemistry of the atmosphere

Chemistry	Trilogy			☑
4.9.1.1–2	5.9.1.1–2	The atmosphere	135	☐
4.9.1.3	5.9.1.3	How oxygen increased	136	☐
4.9.1.4	5.9.1.4	How carbon dioxide decreased	137	☐
4.9.2.1	5.9.2.1	Greenhouse gases	138	☐
4.9.2.2	5.9.2.2	Human activities and greenhouse gases	139	☐
4.9.2.3	5.9.2.3	Global climate change	140	☐
4.9.2.4	5.9.2.4	Carbon footprint	141	☐
4.9.3.1–2	5.9.3.1–2	Atmospheric pollutants from fuels	142	☐
		Examination practice 9	**143**	☐

Topic 10 Using resources

Chemistry	Trilogy		
4.10.1.1	5.10.1.1	Sustainable development	145
4.10.1.2	5.10.1.2	Potable water	146
8.2.8 RPA 8		**Analysing and purifying water samples**	**147**
4.10.1.3	5.10.1.3	Waste water treatment	148
4.10.1.4	5.10.1.4	Alternative methods of extracting metals | **Higher Tier only**	149
4.10.2.1	5.10.2.1	Life cycle assessment	150
4.10.2.2	5.10.2.2	Reuse and recycling	151
4.10.3.1		Corrosion and its prevention | **Chemistry only**	152
4.10.3.2		Alloys as useful materials | **Chemistry only**	154
4.10.3.3		Using polymers | **Chemistry only**	155
4.10.3.3		Clay ceramics and glass | **Chemistry only**	156
4.10.3.3		Composites | **Chemistry only**	157
4.10.4.1		The Haber process 1 | **Chemistry only**	158
4.10.4.1		The Haber process 2 | **Chemistry (Higher Tier only)**	159
4.10.4.2		Production and uses of NPK fertilisers | **Chemistry only**	160
		Examination practice 10	**162**

Examination practice answers	164
Index	173
Notes, doodles and exam dates	177
Levels based mark schemes for extended response questions	178
Command words	179
Key terms in practical work	180
Useful equations	181
Periodic table	182
Examination tips	**183**

MARK ALLOCATIONS

Green mark allocations[1] on answers to in-text questions throughout this guide help to indicate where marks are gained within the answers. A bracketed '1' e.g.[1] = one valid point worthy of a mark. In longer answer questions, a mark is given based on the whole response. In these answers, a tick mark[✓] indicates that a valid point has been made. There are often many more points to make than there are marks available so you have more opportunity to max out your answers than you may think.

TOPICS FOR PAPER 1

Information about Paper 1:

Separate Chemistry 8462:

Written exam: 1 hour 45 minutes
Foundation and Higher Tier
100 marks
50% of the qualification grade
All questions are mandatory

Trilogy 8464:

Written exam: 1 hour 15 minutes
Foundation and Higher Tier
70 marks
16.7% of the qualification grade
All questions are mandatory

Specification coverage

The content for this assessment will be drawn from Topics 1–5 (Topics 8–12 Trilogy): Atomic structure and the periodic table; Bonding, structure, and the properties of matter; Quantitative chemistry; Chemical changes; and Energy changes.

Questions

Multiple-choice, structured, closed short answer and open response questions. They may include calculations.

Questions assess skills, knowledge and understanding of Chemistry.

CHEMISTRY 4.1.1.1 TRILOGY 5.1.1.1

ATOMS, ELEMENTS AND COMPOUNDS

All substances are made from **atoms**. An atom is the smallest part of an **element**.

Elements and compounds

About 100 different elements are known. Each element is represented by a chemical symbol. These consist of a capital letter, which may be followed by one other letter in lower case. For example, C represents carbon and Cr represents chromium. The full **periodic table** (on **page 182**) shows all the known elements but no compounds.

A **compound** is a substance that consists of two or more different elements. The elements in a compound are chemically combined in fixed proportions as a result of a chemical reaction. For example, carbon dioxide always contains one carbon atom for every two oxygen atoms. A **chemical formula** shows the symbol and number of each element in a unit of a compound. The chemical formula of carbon dioxide is always CO_2.

Oxygen is an element, but carbon dioxide and water are compounds

Oxygen   Carbon dioxide   Water

Chemical reactions

Chemical reactions always produce one or more new substances. They involve energy changes that can often be detected. A chemical reaction is the only way to produce a compound from elements, and to separate a compound back into the elements it contains. You can show what happens in a chemical reaction using a **word equation**. This shows the names of the **reactants** (the substances that react together) and the **products** (the substances made in the reaction).

1. Here is the word equation for a chemical reaction.

    silver nitrate + sodium chloride → silver chloride + sodium nitrate

    Describe what this equation shows about the reaction. [2]

2. The chemical formula for water is H_2O. Explain what this shows [4]

    1. Silver nitrate and sodium chloride react[1] to produce silver chloride and sodium nitrate[1].
    2. Water contains hydrogen and oxygen[1] atoms[1] chemically bonded[1] in the ratio 2 : 1[1]

CHEMISTRY 4.1.1.1, 4.2.2.2
TRILOGY 5.1.1.1, 5.2.2.2

CHEMICAL EQUATIONS

Chemical reactions can be represented using **balanced equations**. These use chemical symbols and formulae rather than words.

Formulae with brackets

The formula for aluminium hydroxide, $Al(OH)_3$, represents 1 aluminium, $(3 \times 1) = 3$ oxygen and $(3 \times 1) = 3$ hydrogen.

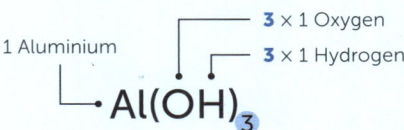

> Do not change chemical formulae in order to balance an equation. For example, H_2O is water but H_2O_2 is hydrogen peroxide.

Ammonium sulfate is commonly used as a soil fertiliser. Its formula $(NH_4)_2SO_4$ represents:

$(2 \times 1) = 2$ nitrogen $(2 \times 4) = 8$ hydrogen   1 sulfur   4 oxygen

Balancing equations

A chemical equation is balanced when the numbers of atoms of each element in the reactants and products are the same. It is important to write the correct symbol or formula for each substance, and to use balancing numbers where necessary.

Methane burns in oxygen to produce carbon dioxide and water:

$$CH_4 + 2O_2 \rightarrow CO_2 + 2H_2O$$

The balancing number 2 in front of the formula for water means that there are:

$(2 \times 2) = 4$ hydrogen atoms and $(2 \times 1) = 2$ oxygen atoms

Check: On each side of the arrow: 1 carbon atom, 4 hydrogen atoms, and $(2 + 2) = 4$ oxygen atoms.

1. Explain why the following chemical equation is balanced: [3]

$$Ca + 2H_2O \rightarrow Ca(OH)_2 + H_2$$

2. Sodium reacts with oxygen. Complete the equation for this reaction. [2]

$$\_\_\_Na + \_\_\_O_2 \rightarrow \_\_\_Na_2O$$

3. Potassium reacts with water to produce hydrogen and potassium hydroxide:

$$2K(s) + 2H_2O(l) \rightarrow H_2(g) + 2KOH(aq)$$

Describe what the state symbols show about this reaction. [4]

State symbols

State symbols show the physical state of a substance in a reaction, or whether it is in aqueous solution (dissolved in water).

1. The 2 in $Ca(OH)_2$ shows that there are 2 oxygen atoms[1] and 2 hydrogen atoms in the formula[1]. On both sides: 1 calcium atom, 4 hydrogen atoms, and 2 oxygen atoms[1].
2. **$4Na + O_2$**[1] **$\rightarrow 2Na_2O$**[1]
3. Potassium is in the solid state[1], water is in the liquid state[1], hydrogen is in the gas state[1] and potassium hydroxide is in aqueous solution.[1]

AQA GCSE **Chemistry 8462 / 8464** – Topic 1

HALF EQUATIONS AND IONIC EQUATIONS

Ions are charged particles formed when atoms, or groups of atoms, lose or gain **electrons**. Half equations and ionic equations are used in reactions involving ions.

Half equations

A **half equation** can show how a substance loses electrons to form ions. An electron is shown as e^- in half equations. The superscript negative sign shows that it carries a single negative charge.

It is important to balance the charges in equations so that there are equal numbers of positive charges and negative charges, as well as equal numbers of atoms and ions.

Sodium atoms lose one electron when sodium reacts with non-metals:
$$Na \rightarrow Na^+ + e^-$$

Calcium atoms lose two electrons when calcium reacts with non-metals:
$$Ca \rightarrow Ca^{2+} + 2e^-$$

Half equations can also show how a substance gains electrons to form ions. For example, chlorine atoms gain electrons when chlorine reacts with metals:
$$Cl_2 + 2e^- \rightarrow 2Cl^-$$

You can also use half equations to show what happens at each electrode during **electrolysis**. During the electrolysis of molten calcium chloride, calcium ions gain electrons to form calcium, and chloride ions lose electrons to form chlorine. The half equations for these reactions are:
$$Ca^{2+} + 2e^- \rightarrow Ca \qquad 2Cl^- \rightarrow Cl_2 + 2e^-$$

---

1. Aluminium reacts with oxygen to produce aluminium oxide. Complete each half equation.
   (a) $Al \rightarrow Al^{3+} + $ ____ [1]
   (b) $O_2 + $ ____ $\rightarrow $ ____ O^{2-} [1]

2. Oxygen forms at the positive electrode during the electrolysis of acidified water.
   Complete the ionic equation for this reaction: ____$OH^- \rightarrow O_2 + $ ____$H_2O + $ ____e^- [2]

1. (a) $Al \rightarrow Al^{3+} + \mathbf{3e^-}$ [1]   (b) $O_2 + \mathbf{4e^-} \rightarrow \mathbf{2}O^{2-}$ [1]
2. $\mathbf{4}OH^- \rightarrow O_2 + \mathbf{2}H_2O + \mathbf{4}e^-$ [2]

Ionic equations

An ionic equation shows how two ions combine to produce a substance. For example, copper(II) ions react with chloride ions to produce copper(II) chloride:
$$Cu^{2+} + 2Cl^- \rightarrow CuCl_2$$

Just like half equations, it is important to balance the charges.

3. Complete this ionic equation: $Pb^{2+}(aq) + $ ____$Br^-(aq) \rightarrow PbBr_2(s)$ [2]

3. $Pb^{2+}(aq) + \mathbf{2}Br^-(aq) \rightarrow PbBr_2(s)$ [1]

Eight elements exist as diatomic molecules (containing two atoms): $H_2, N_2, O_2, F_2, Cl_2, Br_2, I_2, At_2$.

CHEMISTRY 4.1.1.2 | TRILOGY 5.1.1.2

SEPARATING MIXTURES

The substances in a mixture can be separated using processes such as **filtration**.

Mixtures

A **mixture** contains two or more elements or compounds. Unlike compounds, the individual substances in mixtures are not chemically combined. This means that:
- the chemical properties of each substance stay the same
- the substances can be separated from one another by physical processes.

No chemical reactions happen when a mixture is separated, so no new substances form.

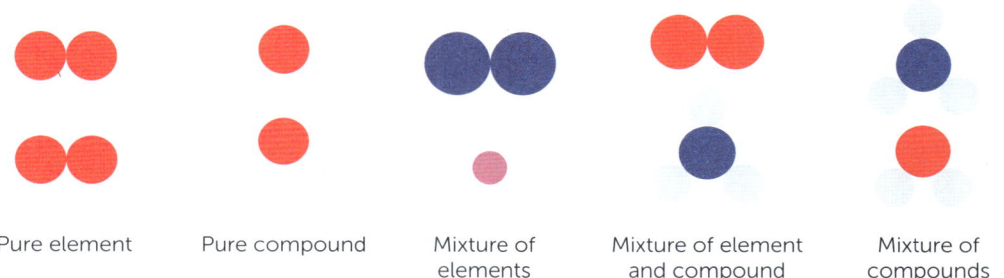

Pure element | Pure compound | Mixture of elements | Mixture of element and compound | Mixture of compounds

Filtration

A **soluble** substance will dissolve in a **solvent** such as water to form a **solution**. An **insoluble** substance will not dissolve. Filtration is a separation method that can separate an insoluble substance from a liquid, gas or solution.

> Explain why sand can be filtered from a mixture of sand and salt solution. [4]
>
> Sand does not dissolve in water[1]. Its particles are too large to pass through the microscopic holes in the filter paper[1]. The dissolved salt particles are small enough to pass through[1] together with the water particles.[1]

Using filtration to separate sand from a mixture of sand and water

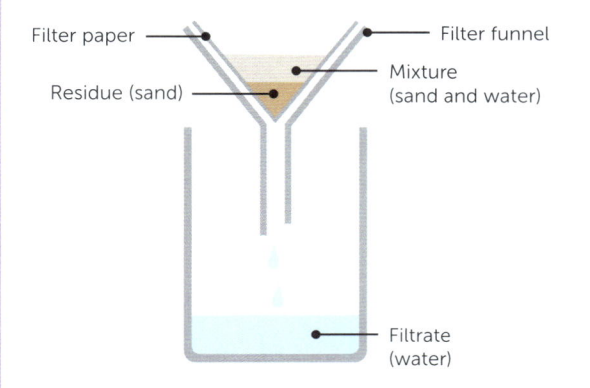

⭐ You can work out the chemical formula of a compound from a diagram that shows its atoms or ions. You can revise how to do this on **page 28**.

CRYSTALLISATION

Crystallisation separates a solid solute from a solution.

Solubility

When no more solute will dissolve in a given volume of solvent, a solution is described as being a **saturated** solution. **Solubility** is the mass of solute in a saturated solution at a given temperature.

The solubility of most substances in the solid state increases as the temperature increases. For example, the solubility of copper(II) sulfate in 100 cm^3 of water is 32 g at 20 °C, but 84 g at 80 °C. If you cool 100 cm^3 of a saturated solution from 80 °C to 20 °C, (84 − 32) = 52 g of copper(II) sulfate cannot stay dissolved and will return to the solid state. This is how crystallisation works.

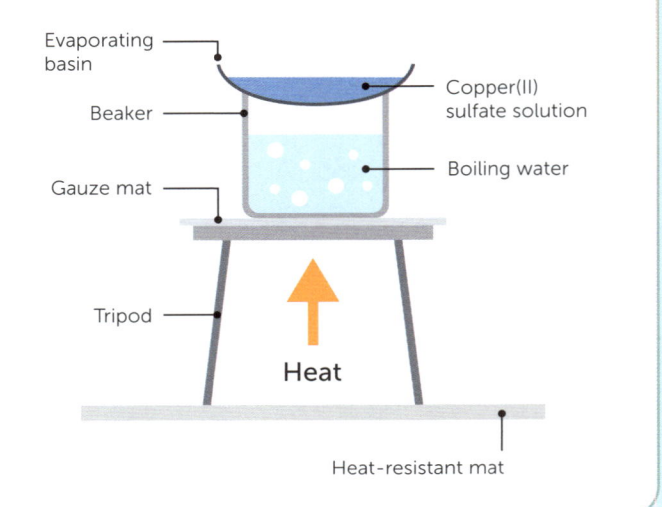

Crystallising copper(II) sulfate over a boiling water bath

Crystallisation method

(a) Put the solution into an evaporating basin and heat over a boiling water bath. The solution becomes more concentrated as water in the solution evaporates.

(b) Stop heating before all the water has evaporated. Crystals form as the solution cools down.

(c) Leave the evaporating basin with its contents aside for a few days, such as on a windowsill. Remove the crystals and gently pat them dry with a paper towel or filter paper.

1. Explain why a boiling water bath is used instead of heating directly with a Bunsen burner. [3]
2. Suggest **one** way to dry the crystals other than leaving the evaporating basin for a few days. [1]

1. A boiling water bath heats the solution more gently.[1] This reduces the chance of hot solids or liquids jumping out of the evaporating basin[1], which would be unsafe[1].
2. Place the evaporating basin in a warm oven.[1]

CHROMATOGRAPHY

Chromatography separates a mixture of coloured solutes in a solution.

Paper chromatography

Chromatography relies on two 'phases':
- a **stationary phase** that does not move - usually a porous solid.
- a **mobile phase** that moves through the stationary phase.

In paper chromatography, the stationary phase is contained in the paper, and the mobile phase is a solvent such as water or propanone.

Different solutes in a solution form chemical bonds with both phases. The relative strengths of these bonds determine how far a solute travels up the paper with the solvent. The more strongly a solute bonds to the mobile phase, the further it travels up the paper.

Paper chromatography of a sample of ink

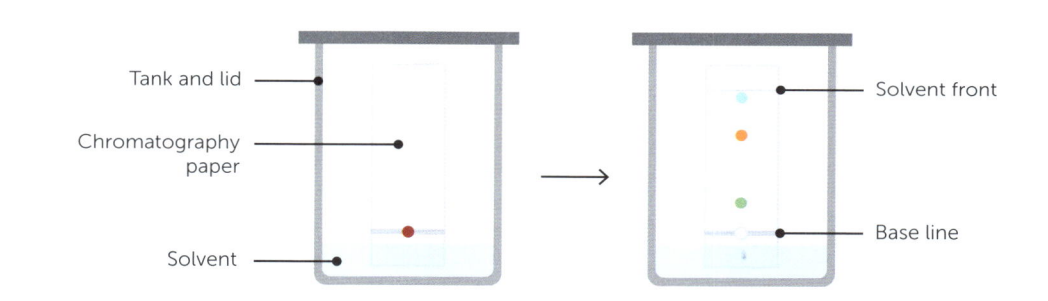

Method

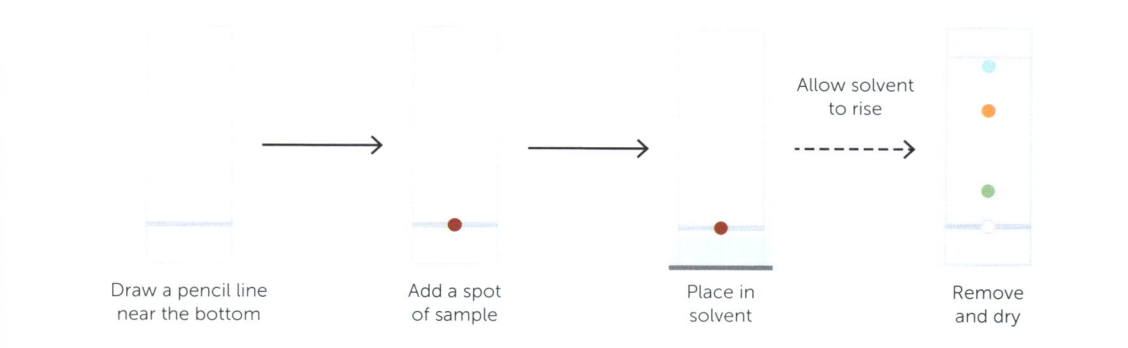

1. Explain why the sample must be higher than the solvent in paper chromatography. [2]
2. Give **two** ways to determine whether two coloured spots are the same substance. [2]

   1. To stop it dissolving in the solvent[1] so it does not leave the chromatography paper[1].
   2. They will be same colour[1] and travel the same distance on the same chromatogram[1].

| 4.1.1.2 | 5.1.1.2 |

SIMPLE DISTILLATION

Simple distillation separates the solvent from a solution.

Boiling points

Distillation relies on the different **boiling points** of the components in a solution. In a solution made by dissolving a solid in a liquid, the **solvent** boils at a lower temperature than the **solute**. This means that when the solution is heated enough:
- solvent evaporates and escapes the solution
- the solute is left behind, and the solution gradually becomes more concentrated.

State changes

In simple distillation, solvent evaporates and travels as a vapour into a **condenser**. This is a glass tube with cold water surrounding it. The vapour cools and condenses inside the condenser. Pure liquid solvent is collected as a **distillate** as it leaves the condenser.

1. Give **one** practical use of simple distillation. [1]
2. During the simple distillation of pale blue ink, the colour gradually turns dark blue.
   Explain this observation. [2]

   1. Drinking water can be made by the simple distillation of sea water[1].
   2. Blue pigment stays in the ink as the solvent leaves[1] so the ink gets more concentrated[1].

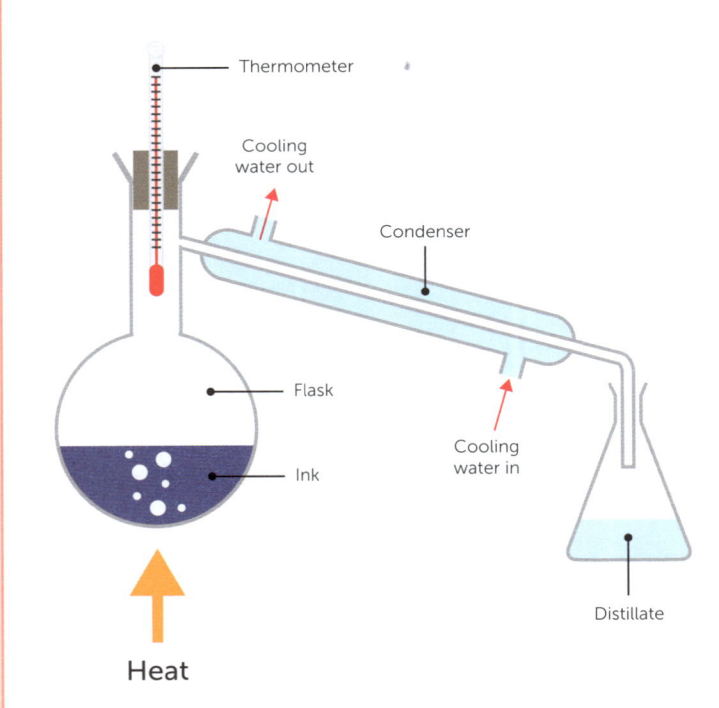

Simple distillation of blue ink

8 ClearRevise

4.1.1.2   5.1.1.2

FRACTIONAL DISTILLATION

Fractional distillation separates a liquid from a mixture of liquids.

How it works

Fractional distillation works in a similar way to simple distillation, but the apparatus includes a **fractionating column**. This is placed between the flask and the condenser.

During fractional distillation, a **temperature gradient** forms:
- The bottom of the fractionating column becomes hotter than the top.
- When the mixture of liquids is heated, the liquid with the lowest boiling point evaporates first. Its vapour travels up the fractionating column and into the condenser, where it cools and condenses.
- The liquid which leaves the condenser is called a **fraction** because it is only a part of the original mixture of liquids.
- With continued heating, liquids with higher boiling points may be collected, one after the other.

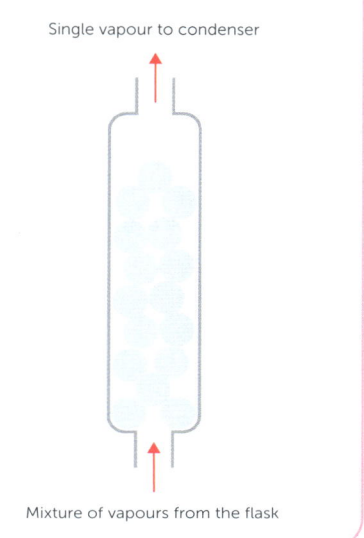

Single vapour to condenser

Mixture of vapours from the flask

Choosing separation and purification methods

Different separation methods are suitable for different mixtures. They can be used on their own, or in a combination of two or more different methods.

Method	Separates
Filtration	Insoluble solid from a liquid
Crystallisation	Solid solute from a solution
Simple distillation	Solvent from a solution
Fractional distillation	Liquid from a mixture of liquids
Chromatography	Different coloured solutes in a solution

1. Give **two** mixtures that can be separated using fractional distillation. [2]
2. Describe the function of the fractionating column in fractional distillation. [3]
3. Describe how to separate sand and salt from a mixture of sand and salt solution. [3]

1. Crude oil[1] and the mixture formed by the fermentation of sugar[1].
2. The column provides a large surface area[1] for vapours to cool, condense and then evaporate again[1]. This improves the separation of the different liquids in a mixture[1].
3. Filter to separate the sand[1] then use crystallisation to produce salt[1] from the filtrate[1].

DEVELOPING THE ATOMIC MODEL

The **atomic model** has changed over time because of new experimental evidence.

Atomic theories

In the early 19th century, atoms were imagined as tiny, solid spheres. The discovery of the **electron** by J.J. Thomson in 1897 led to his **plum pudding model**. This model was disproved by a series of results from the **alpha particle scattering experiment**.

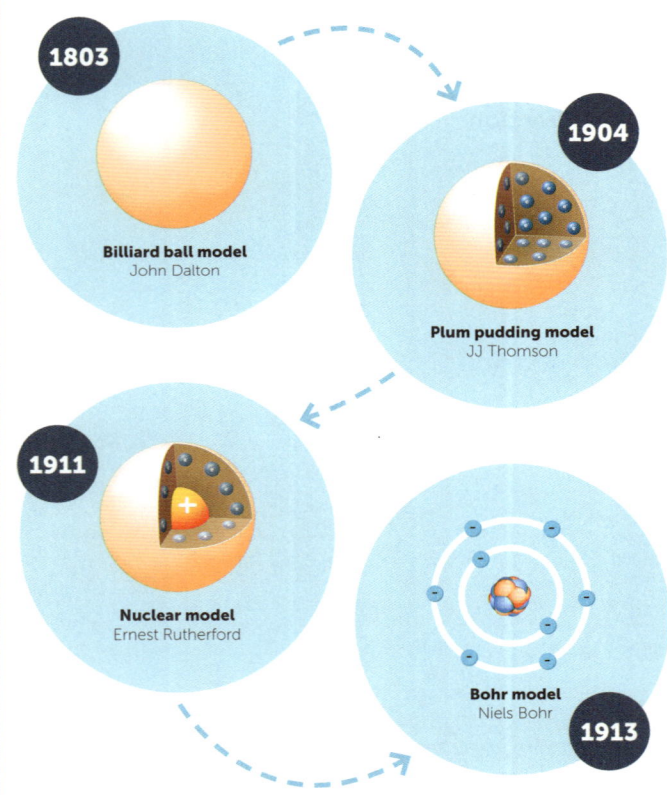

In this experiment, beams of alpha particles (tiny positively charged particles) were aimed at thin gold foil. The results led to the **nuclear model**. Shortly afterwards, Niels Bohr carried out theoretical calculations showing that electrons orbit the nucleus at set distances. Observations from experiments supported his **electron shell model** and also showed the existence of positively charged **protons**. About 20 years later, James Chadwick demonstrated the existence of **neutrons**.

---

1. Compare the plum pudding and nuclear models of the atom [3]
2. The diagram shows paths taken by alpha particles through gold foil.

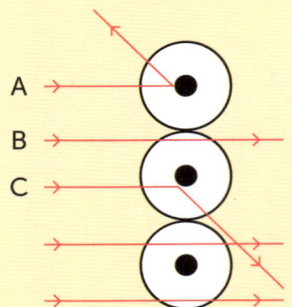

(a) Give a reason why most particles followed the path labelled B. [1]
(b) Explain why some particles followed path C. [2]
(c) Explain why a very small number of particles followed path A. [2]

---

1. Both models have negatively charged electrons[1]. These are embedded in a sphere of positive charge in the plum pudding model[1] but surround a positively charged nucleus in the nuclear model[1].
2. (a) Atoms are mostly empty space[1] so particles passed straight through.
   (b) The nucleus was positively charged[1] so it repelled the positively charged alpha particles[1].
   (c) The nucleus was very small[1] but it had a relatively high mass[1] and high charge[1].

SUBATOMIC PARTICLES

Atoms are very small, but they are made from even smaller **subatomic particles**.

Protons, neutrons and electrons

The nuclei of all atoms contain protons, and almost all contain neutrons. Electrons are arranged around the nucleus. These three subatomic particles have different masses and electrical charges. Instead of giving the actual values, mass and charge are compared to the mass and charge of a proton. This gives relative values.

Name of particle	Relative mass	Relative charge
Proton	1	+1
Neutron	1	0
Electron	Very small	−1

Protons and neutrons have a lot more mass than electrons. This means that most of an atom's mass is in its nucleus. An atom is neutral overall because it contains the same number of protons and electrons.

Atomic number

An **atomic number** is the number of protons in an atom:

- the atoms of an element have the same atomic number
- the atoms of different elements have different atomic numbers.

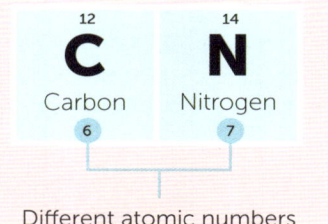

Different atomic numbers

Atomic radius

Atoms vary in size but all of them are very small. Their typical radius is about 0.1 nm. Remember that 1 nm is 10^{-9} m in standard form, so this is 1×10^{-10} m. The radius of a nucleus is about 1×10^{-14} m, around 10 000 times smaller than the atom itself.

The actual mass and charge of a proton are incredibly small, just 1.67×10^{-27} kg and $+1.60 \times 10^{-19}$ C. This is why relative values are used.

The mass of the electron is 1836 times less than the mass of proton.

---

1. A bacterium is 2×10^{-6} m long. Calculate how many times larger the bacterium is compared to an atom. [2]
2. Compare the charges of protons, neutrons and electrons. [2]

   1. Diameter of atom = $2 \times 1 \times 10^{-10}$ m[1] Number of times larger = $\dfrac{2 \times 10^{-6} \text{ m}}{2 \times 10^{-6} \text{ m}}$ = 10,000 times[1].

   2. Protons and electrons are charged but neutrons are not[1]. Protons and electrons have opposite charges[1].

AQA GCSE **Chemistry** 8462 / 8464 – Topic 1

4.1.1.5 **5.1.1.5**

ISOTOPES

Isotopes are atoms of the same element with different atomic mass numbers.

Representing atoms

The **mass number** of an atom is the total number of protons and neutrons in its nucleus. Atoms are represented using their mass number, atomic number and chemical symbol.

Mass number → $^{39}_{19}$K
Atomic number →

The isotopes of an element have identical chemical properties because their atoms have the same number of electrons. You can revise the relationship between electrons and properties on **page 15**.

Numbers of particles in atoms

The numbers of neutrons, protons and electrons in an atom are calculated from its mass number and atomic number:
- number of neutrons = mass number – atomic number
- number of protons = atomic number

Remember that atoms have equal numbers of protons and electrons. This means that an atom of $^{39}_{19}$K contains (39 – 19) = 20 neutrons, 19 protons and therefore 19 electrons.

Numbers of particles in ions

An **ion** is an atom or molecule with a net electrical charge greater or less than 0. It forms when an atom loses or gains electrons. Positively charged ions contain fewer electrons than protons, and negatively charged ions contain more electrons than protons. The number of charges on an ion, and whether it is positively charged or negatively charged, is shown using a superscript:
- K^+ is an ion carrying a single positive charge
- O^{2-} is an ion carrying two negative charges.

---

1. Explain why $^{23}_{11}$X and $^{23}_{12}$X are **not** isotopes of the same element. [2]
2. Explain, in terms of subatomic particles, why $^{12}_{6}$C and $^{14}_{6}$C are isotopes of an element. [3]
3. Give the numbers of protons, neutrons and electrons in the following ions. [6]
    (a) $^{39}_{19}K^+$ (b) $^{18}_{8}O^{2-}$

    1. All atoms of an element have the same atomic number[1] but these have different atomic numbers[1].
    2. Both contain 6 protons[1] but different numbers of neutrons[1]. ^{12}C atoms have (12 – 6) = 6 neutrons and ^{14}C atoms have (14 – 6) = 8 neutrons[1].
    3. (a) 19 protons[1], (39 – 19) = 20 neutrons[1], (19 – 1) = 18 electrons[1]
       (b) 8 protons[1], (18 – 8) = 10 neutrons[1], (8 + 2) = 10 electrons[1]

4.1.1.6 5.1.1.6

RELATIVE ATOMIC MASS

Periodic tables often show the **relative atomic mass** of each element.

A weighted average

Mass number and relative atomic mass are not the same thing, even though they may have the same value for some elements. Mass number refers to atoms and relative atomic mass refers to elements:
- mass numbers are always whole numbers
- relative atomic masses are rarely whole numbers, unless rounded up or down.

Relative atomic mass (symbol A_r) is an average value that takes into account the relative **abundances** of all the isotopes in a sample of an element.

Calculating relative atomic masses

In a typical sample of chlorine, 75% of the atoms are ^{35}Cl and 25% are ^{37}Cl:

Relative atomic mass = $\dfrac{(75 \times 35) + (25 \times 37)}{(75 + 25)} = \dfrac{2625 + 925}{100} = \dfrac{3550}{100}$

Relative atomic masses have no units. Where a periodic table shows the relative atomic mass of each element, take care not to confuse these values with mass numbers. The box for chlorine in the periodic table shows 35.5 rather than 35 or 37.

A_r values are given relative to the mass of a ^{12}C atom, which is taken as 12 exactly.

---

1. Give **one** reason why 35.5 cannot be the mass number of chlorine. Answer in terms of subatomic particles. [2]
2. The table shows the percentage abundance of three zinc isotopes. Calculate the relative atomic mass of zinc. Give your answer to **1 decimal place**. [3]

Isotope	Percentage abundance (%)
^{64}Zn	51
^{66}Zn	29
^{68}Zn	20

1. If it was 35.5 it would mean that chlorine atoms could not have whole numbers of protons or neutrons[1].

2. An answer of 65.4 scores **3** marks.

   relative atomic mass = $\dfrac{(51 \times 64) + (29 \times 66) + (20 \times 68)}{(51+29+20)}$ [1]

   = $\dfrac{3264 + 1914 + 1360}{100} = \dfrac{6538}{100}$ [1]

   = 65.4 to 1 decimal place[1].

AQA GCSE Chemistry 8462 / 8464 – Topic 1

4.1.1.7　5.1.1.7

ELECTRONIC STRUCTURE

The **electronic structure** of an atom shows how its electrons are arranged around its nucleus.

Energy levels

Electrons can occupy **energy levels** in atoms. The energy of each energy level increases as the distance from the nucleus increases. Each energy level can hold different numbers of electrons. The table shows the maximum numbers of electrons in each energy level of the first 20 elements in the periodic table.

Energy level	Maximum number of electrons
1	2
2	8
3	8
4	18

Energy levels may also be described as electron **shells**. Electrons in an atom occupy and fill the lowest available energy levels, or the innermost available shells, before any outer shells.

⭐ You can easily check your electronic structures in exams using the periodic table:

- total number of electrons = atomic number
- number of levels = period number
- number of electrons in outer level = group number (except for Group 0)

Deducing electronic structures

Electronic structures can be represented by numbers and by diagrams. The atomic number of calcium is 20, so its atoms have 20 electrons. The electronic structure of calcium therefore, is 2,8,8,2. This shows that:
- 2 electrons occupy the first energy level
- 8 electrons occupy the second and third energy levels, leaving
- 2 electrons to occupy the fourth energy level.

The diagram for this electronic structure has four concentric circles, one for each energy level. Each electron is shown as a cross. You can spread the crosses evenly around each circle, but they are easier to count if you show pairs of crosses (as here). This also helps you to show molecules.

An electron diagram for calcium

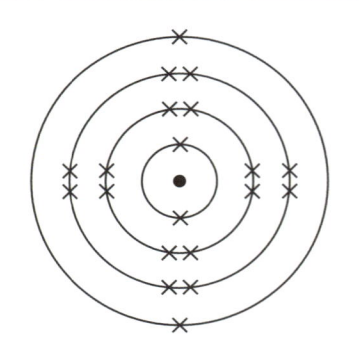

Using the Period table, write the electronic structures of the following atoms. [2]

(a) Lithium　　(b) Carbon　　(c) Aluminium　　(d) Sulfur　　(e) Argon

(a) 2,1 [1]　　(b) 2,6 [1]　　(c) 2,8,3 [1]　　(d) 2,8,6 [1]　　(e) 2,8,8 [1]

THE PERIODIC TABLE

The periodic table shows the elements in order of increasing atomic number.

Groups

The rows in the periodic table are called periods. The columns in the periodic table are called groups. The elements in a group have these properties:
- similar chemical reactions, and
- the same number of electrons in their highest occupied energy level (outer shell).

The number of outer electrons is the same as the group number. For example, all the elements in Group 1 have 1 outer electron and all the elements in Group 7 have 7 outer electrons. All the elements in Group 0 have full outer shells.

The number of outer electrons determine the chemical properties of an element, which is why elements in a group have similar chemical properties.

Elements and their positions

An element's chemical properties include its **reactivity** and reactions. You predict an element's typical chemical properties from its atomic number shown in the periodic table. There is a copy of the periodic table on **page 182**.

An electron diagram for oxygen

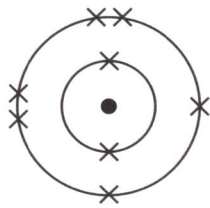

The electronic configuration and electron diagram for oxygen shows that it has two occupied energy levels. This means that oxygen is placed in period 2 of the periodic table.

1. Nitrogen is placed in period 2, group 5, of the periodic table. Give the electronic structure of nitrogen and its atomic number. [2]
2. The atomic number of sodium is 11. Explain the position of sodium in the periodic table. [3]

   1. 2,5[1] atomic number (2 + 5) = 7[1].
   2. The electronic structure is 2,8,1[1], so sodium is in period 3[1] Group 1[1].

You can revise the general properties of metals and non-metals, and their positions on the periodic table, on **page 17**.

The complete table is printed on page 182.

THE DEVELOPMENT OF THE PERIODIC TABLE

New scientific discoveries and ideas have led to the modern periodic table.

Mendeleev's work

Mendeleev first placed the elements in order of increasing atomic weight, as other scientists had done. However, he left gaps in his periodic table for elements that he thought would be discovered later. This meant that:
- elements with similar chemical properties were placed in groups
- he could make predictions about the **physical properties** and **chemical properties** of the unknown elements.

Mendeleev sometimes changed the positions of some elements to match their properties better. However, he was unable to adequately explain why this should work.

The positions of tellurium and iodine are an example of one of Mendeleev's 'pair reversals'

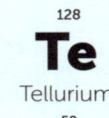 

1. Give **one** reason why Mendeleev's predictions were important to the success of his table. [1]
2. Explain how scientists today can explain why Mendeleev was correct when he changed the order of some elements in his periodic table. [4]

   1. When the missing elements were discovered, their properties were found to be similar to his predictions[1].
   2. Protons, neutrons and electrons had not been discovered then[1] and Mendeleev did not know about isotopes[1]. The modern periodic table places elements in order of atomic number[1] but the existence of isotopes means that some elements have higher relative atomic masses than expected from their atomic number alone[1].

Atomic weights

Early attempts to produce periodic tables relied on **atomic weights**. These were first determined using attempts at working out chemical formulae but were often inaccurate. They differed in value from modern relative atomic masses, and were often too high or too low.

Early periodic tables were incomplete because a lot of elements were unknown at the time. When a strict order of increasing atomic weight was used, some elements were located in the wrong groups. Dmitri Mendeleev devised periodic tables that overcame some of the problems with other early periodic tables.

⭐ The periodic table given to you in exams shows three other pair reversals, including argon and potassium. See if you can find the other two on **page 182**.

METALS AND NON-METALS

Most elements are metals rather than non-metals.

Chemical properties

An element is a **metal** if it forms positively charged **ions**, and a **non-metal** if it does **not** form positively charged ions. Metals and non-metals are found in different places on the periodic table.

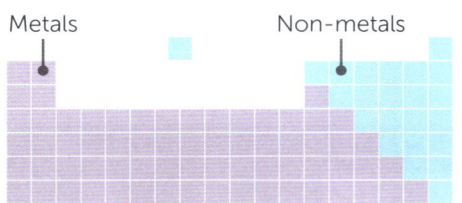

Chemical bonding

Metals and non-metals form different types of chemical **bonds**:

- metals have metallic bonding
- non-metal elements have covalent bonding
- compounds of metals and non-metals have ionic bonding.

See bonds on **page 26**.

The elements in groups 1 and 2, and between groups 2 and 3, are all metals.

Physical properties

In general, metals and non-metals have opposite physical properties. For example, metals are malleable – they can be hammered into shape without shattering. Non-metals in the solid state are brittle – they shatter when hammered. The table compares some other typical properties.

Property	Metals	Non-metals
Appearance	Shiny	Dull
Melting and boiling points	High	Low
Density	High	Low
Ability to conduct electricity and thermal energy	Good	Poor

1. Describe the positions of the metals and non-metals in the periodic table. [1]
2. The element mercury is in the liquid state at room temperature. Give a reason why this is unusual. [1]
3. Hydrogen is a non-metal. It can form H$^+$ ions and H$^-$ ions. Explain why this is unusual. [3]

   1. Metals are found towards the bottom and left, and non-metals are found towards the top and right[1].
   2. Metals are usually in the solid state at room temperature[1].
   3. Elements that do not form positive ions are non-metals[1] but hydrogen can form positive ions just like metals do[1].

AQA GCSE **Chemistry 8462 / 8464 – Topic 1**

GROUP 0

The Group 0 elements are called the **noble gases**.

Chemical properties

The Group 0 elements are unreactive non-metals. The highest occupied energy levels of their atoms are completely filled. These stable arrangements mean that the noble gases:

- have little tendency to lose or gain electrons in chemical reactions, so they do not easily form **ionic compounds**.
- have little tendency to share electrons, so they do not easily form **molecules**.

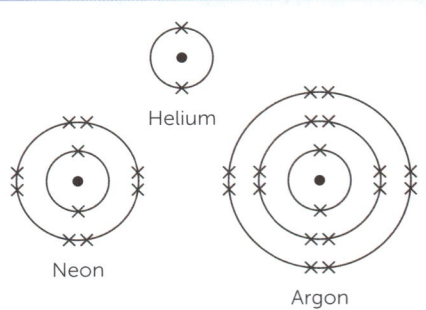

Helium

Neon

Argon

Physical properties

The noble gases have very low boiling points, so they are all in the gas state at room temperature. There is a gradual change or **trend** in their boiling points going down the group.

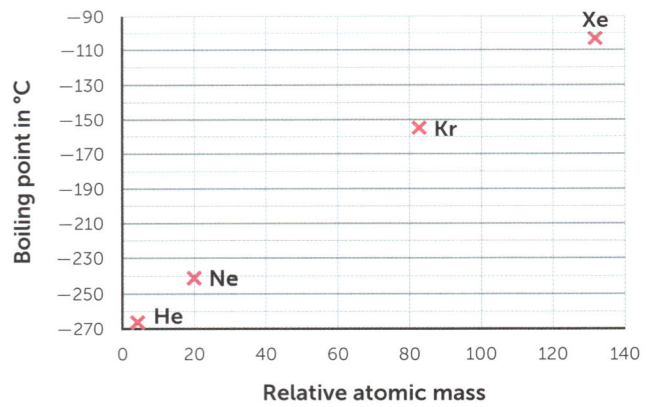

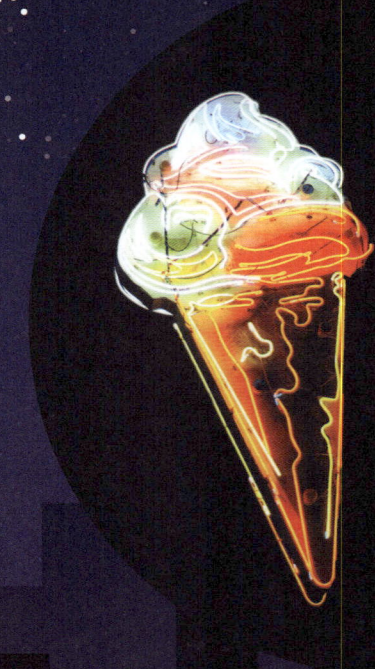

1. Compare the electronic configuration of helium with the electronic configurations of the other elements in Group 0. [1]
2. (a) Describe the relationship in Group 0 between boiling point and relative atomic mass. [1]

   (b) The relative atomic mass of argon is 40. Predict the boiling point of argon. [1]

   1. Helium has 2 outer electrons but the other elements have 8 outer electrons[1].

   2. (a) Boiling point increases as the relative atomic mass increases[1].

   (b) Between −230 °C and −210 °C [1].

Group 0 atoms do lose electrons to form ions when high voltages are applied to them. They give off coloured light when the electrons return to the ions. This is how neon lights work.

GROUP 1

The Group 1 elements are called the **alkali metals**. They have one outer electron.

Reactions with chlorine

The alkali metals react vigorously with chlorine when heated. They burn with coloured flames to produce metal chlorides. In general, where M stands for the alkali metal:

$$2M(s) + Cl_2(g) \rightarrow 2MCl(s)$$

Lithium burns with a red flame, sodium burns with an orange flame, and potassium burns with a lilac flame. The reaction of potassium with chlorine is the most vigorous.

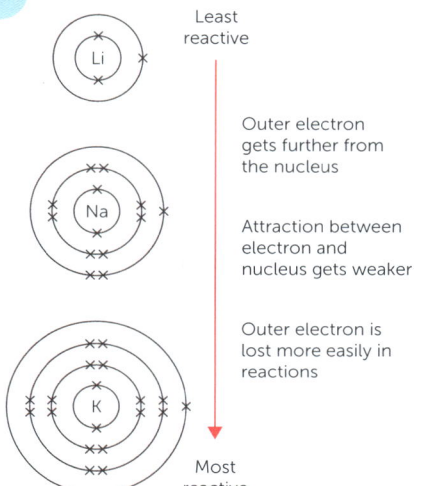

Least reactive — Li
Outer electron gets further from the nucleus
Attraction between electron and nucleus gets weaker
Outer electron is lost more easily in reactions
Most reactive

Lithium-ion batteries are essential components of devices such as smartphones. They are lightweight, rechargeable and store a lot of energy.

Reactivity

Alkali metals lose their outer electrons in reactions with non-metals. They become more reactive going down the group.

Reactions with water

The alkali metals react with water to produce metal hydroxides and hydrogen. In general, where M stands for the alkali metal:

$$2M(s) + 2H_2O(l) \rightarrow 2MOH(aq) + H_2(g)$$

The first three elements all float on water and disappear as they react:
- lithium fizzes steadily
- sodium melts to form a silvery ball, fizzes quickly, and may burn with an orange flame.
- potassium burns with a lilac flame very quickly, then pops or explodes.

The metal hydroxides produced in the reaction are all soluble. They dissolve in water to form **alkaline solutions** that turn **universal indicator** blue or purple.

---

1. (a) Describe the reactions of lithium, sodium and potassium with oxygen. [4]

   (b) Balance the equation for the reaction of lithium with oxygen. Include state symbols.

   ___Li(___) + O_2(___) → ___Li_2O(___) [2]

2. Rubidium is placed below potassium in Group 1. Predict **two** observations you would see when rubidium is added to water. [2]

---

1. (a) Lithium burns with a red flame[1], sodium with an orange flame[1] and potassium with a lilac flame[1]. The reactions become more vigorous from lithium to potassium[1].

   (b) **4**Li(s) + O_2(g) → **2**Li_2O(s) Correctly balanced[1], correct state symbols[1].

2. Two from: a violent or explosive reaction[1], a coloured flame[1], the metal disappears faster than potassium does[1], the water turns universal indicator blue or purple[1].

AQA GCSE Chemistry 8462 / 8464 – Topic 1

GROUP 7

4.1.2.6 5.1.2.6

The Group 7 elements are called the **halogens**. They have seven outer electrons.

Physical properties

The melting points and boiling points of the halogens increase going down the group. Fluorine and chlorine are in the gas state at room temperature, bromine is in the liquid state and iodine is in the solid state. Bromine and iodine slowly produce coloured vapours at room temperature.

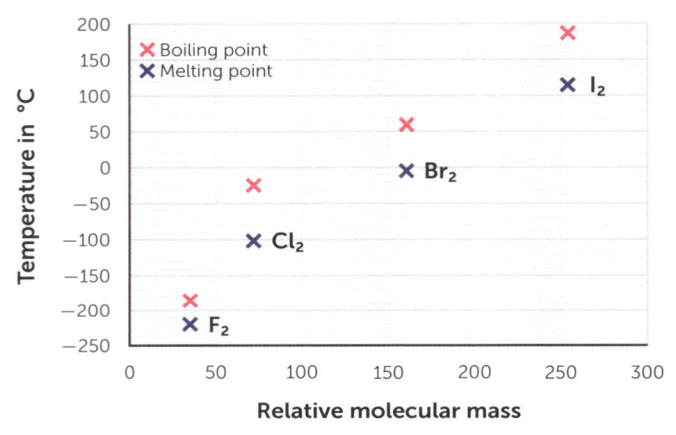

Reactions with metals and non-metals

Halogen molecules each consist of two atoms joined by a **covalent bond**. Halogens react with metals to form **ionic compounds**. For example, chlorine reacts with sodium to produce sodium chloride:

$$2Na(s) + Cl_2(g) \rightarrow 2NaCl(s)$$

Depending on the conditions, halogens can react with some non-metals to form **covalent compounds**. They dissolve in water and react with it to form **acidic solutions**.

Trend in reactivity

Halogens gain electrons in chemical reactions with metals. They become less reactive down the group.

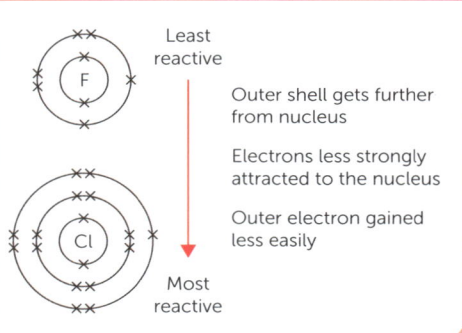

The reaction between chlorine and water creates a solution that is used to kill harmful microorganisms in swimming pools.

Astatine, At, is placed below iodine in Group 7.

(a) Predict the state of astatine at room temperature. Give a reason for your answer. [2]

(b) Predict the formula of astatine. [1]

(a) Solid[1] because iodine is solid and melting points increase down group 7[1]

(b) At_2[1].

4.1.2.6 | 5.1.2.6

HALOGEN DISPLACEMENT REACTIONS

A **displacement reaction** happens when an atom or ion replaces an existing atom or ion in a compound.

Representing displacement reactivity

A more reactive halogen can **displace** a less reactive halogen from its compounds, particularly salts in aqueous solution. For example, chlorine is more reactive than iodine. Chlorine can displace iodine from potassium iodide solution:

$$Cl_2(aq) + 2KI(aq) \rightarrow 2KCl(aq) + I_2(aq)$$

Higher Tier only

Two **half equations** represent this reaction:
- chlorine is **reduced** to chloride ions: $Cl_2(aq) + 2e^- \rightarrow 2Cl^-(aq)$
- iodide ions are **oxidised** to iodine: $2I^-(aq) \rightarrow I_2(aq) + 2e^-$

An **ionic equation** also represents it: $Cl_2(aq) + 2I^-(aq) \rightarrow 2Cl^-(aq) + I_2(aq)$

The potassium ions in the reaction mixture are **spectator ions**. They do not take part in the reaction and are unchanged, which is why they do not appear in the ionic equation.

Determining reactivity

Chlorine, bromine and iodine dissolve in water to form aqueous solutions. These solutions can be mixed with aqueous solutions of **halide** salts, such as sodium chloride solution. A colour change to a darker colour after mixing indicates that a reaction has happened.

A student carries an investigation to determine the reactivity of three halogens. Nine different mixtures of halogen solution and salt solution are prepared. The table shows the results. A tick shows that a visible reaction occurred.

	Potassium Chloride	Potassium Bromide	Potassium Iodide
Chlorine	✗	✓	✓
Bromine	✗	✗	✓
Iodine	✗	✗	✗

(a) Determine the order of reactivity, starting with the most reactive halogen. [1]
(b) Explain your answer to part (a). [3]

(a) Chlorine, bromine, iodine[1].

(b) Chlorine displaces bromine and iodine from their salts[1], bromine displaces iodine but not chlorine[1] and iodine does not displace chlorine or bromine[1].

AQA GCSE Chemistry 8462 / 8464 – Topic 1

TRANSITION METALS VS GROUP 1

The **transition metals** are a block of elements placed between groups 2 and 3 in the periodic table. You need to know about six transition metals in particular.

Reactivity of transition metals

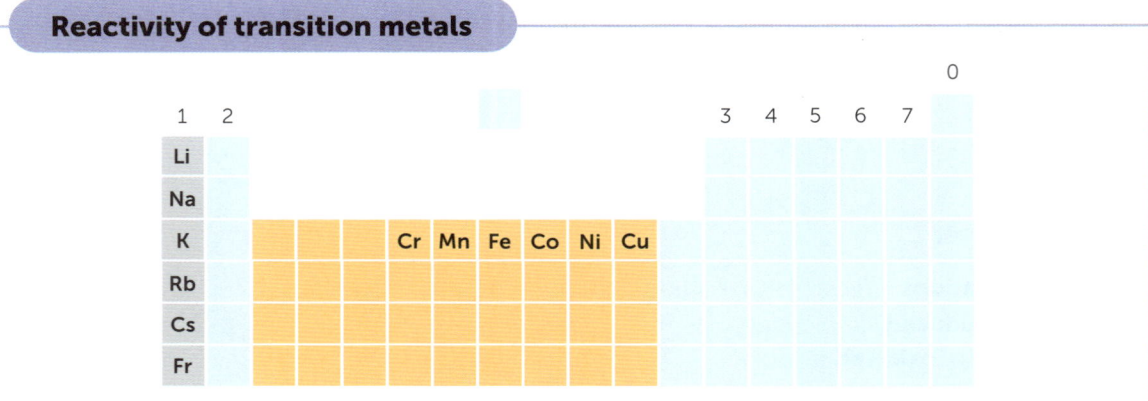

The alkali metals in Group 1 react vigorously with water and oxygen. However, the transition metals:
- do not react with water
- do not react with oxygen unless they are powdered and heated.

The alkali metals and transition metals react with halogens to form metal halides. For example, hot iron reacts with chlorine to form orange-brown iron(III) chloride:

$$2Fe(s) + 3Cl_2(g) \rightarrow 2FeCl_3(s)$$

Physical properties of transition metals

The typical physical properties of the transition metals are opposite to those of the alkali metals.

Property	Group 1 metals	Transition metals
Melting point	Low	High
Density	Low	High
Strength	Low	High
Hardness	Soft	Hard

---

1. Explain why the reaction of iron with water and oxygen is unexpected, given the typical properties of transition metals. [3]
2. Iron powder burns in oxygen to produce iron(III) oxide. Balance the equation for this reaction.
   ___Fe + ___O_2 → ___Fe_2O_3 [1]

   1. Transition metals do not easily react with water or oxygen[1] but iron readily reacts with water in the presence of oxygen[1] to form rust[1].
   2. $4Fe + 3O_2 \rightarrow 2Fe_2O_3$ Correctly balanced.[1]

4.1.3.2 Chemistry

CHARGES, COLOURS, AND CATALYSTS

Many transition metals form ions with different charges, form coloured compounds, and act as catalysts.

Ions

Like all metals, transition metals form positively charged **ions** in chemical reactions. Group 1 metals only form ions that have a single positive charge, such as Na^+ and K^+. Transition metals can form positively charged ions with different numbers of charges.

Transition metals often form ions with two or three positive charges. Roman numbers in brackets distinguish between ions of the same metal but with different charges:
- Fe^{2+} is the iron(II) ion
- Fe^{3+} is the iron(III) ion.

> Manganese(IV) oxide is a black powder that catalyses the breakdown of hydrogen peroxide to water and oxygen.
>
> You can revise catalysts on **page 94**.

Coloured compounds

All alkali metal compounds are white or colourless, but many transition metal compounds are coloured. Some colours are familiar, such as blue copper(II) sulfate and purple potassium manganate(VII), but it is very difficult to predict the colour of an individual compound.

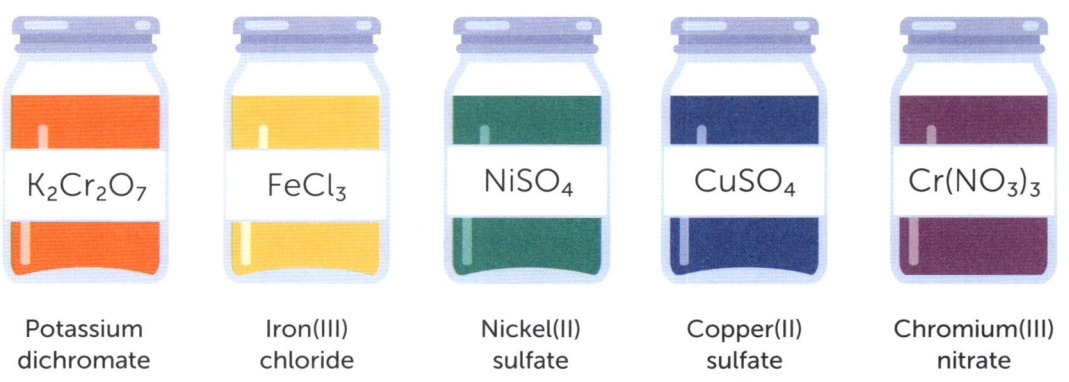

| $K_2Cr_2O_7$ | $FeCl_3$ | $NiSO_4$ | $CuSO_4$ | $Cr(NO_3)_3$ |
| Potassium dichromate | Iron(III) chloride | Nickel(II) sulfate | Copper(II) sulfate | Chromium(III) nitrate |

---

1. (a) Describe what a catalyst is. [2]
   (b) Name the catalyst used in the manufacture of ammonia by the Haber process. [1]
2. Give a reason why iron(II) chloride solution is pale green but iron(III) chloride solution is orange-brown. [3]

> 1. (a) A substance that changes the rate of a chemical reaction[1] but is not used up during the reaction[1].
>   (b) Iron.[1]
> 2. They contain different iron ions[1], Fe^{2+} and Fe^{3+}[1].

AQA GCSE **Chemistry** 8462 / 8464 – Topic 1

TOPIC 1

EXAMINATION PRACTICE

01 Give **two** differences between a compound and a mixture. [2]

02 Iron(III) oxide reacts with carbon to make iron and carbon dioxide.
Balance the equation for this reaction. [1]

$$\_\_\_\_Fe_2O_3 + \_\_\_\_C \rightarrow \_\_\_\_Fe + \_\_\_\_CO_2$$

03 This question is about atoms and ions.
   03.1 Explain how the discovery of the electron led to an improvement in the model of the atom. [2]
   03.2 Complete the following table. [3]

Name of subatomic particle	Relative charge	Relative mass
		1
		Very small
	+1	

   03.3 A nitride ion is represented as $^{15}_{7}N^{3-}$.
Give the numbers of protons, neutrons and electrons in this ion. [3]
   03.4 The diameter of a gold atom is 0.28 nm. Give this value in m in standard form. [1]

04 A sample of copper contains two isotopes, 69% $^{63}_{29}Cu$ and 31% $^{65}_{29}Cu$.
   04.1 Calculate the relative atomic mass of copper in this sample. Give your answer to 3 significant figures. [3]
   04.2 Explain why these isotopes have the same chemical properties. [2]

05 This question is about the periodic table.
   05.1 Describe **two** ways in which Mendeleev overcame some of the problems of earlier periodic tables. [2]
   05.2 The electronic configuration of an element X is 2,8,5.
Determine the position of element X in the periodic table. [2]
   05.3 Neodymium is an element placed between groups 2 and 3. Predict whether neodymium forms negatively charged ion or positively charged ions in reactions. Explain your answer. [2]

06 A dark green ink consists of insoluble carbon particles, mixed with three different coloured substances dissolved in a mixture of two liquids, water and propanol.
Design an experiment to separate all six substances in the ink. Include essential steps and safety precautions in your answer. [6]

07 This question is about Group 0 elements.

07.1 Explain why the elements in Group 0 exist as single atoms, rather than as molecules. [2]

07.2 Describe the trend in boiling points of the Group 0 elements. [1]

The table shows the densities of some Group 0 elements.

Element	He	Ne	Ar	Kr	Xe	Rn
Relative atomic mass	4	20	40	84	131	222
Density in g/dm³	0.179	0.900	1.784		5.89	9.73

07.3 Predict the density of krypton. Explain your answer. [3]

07.4 Oganesson is a Group 0 element that was discovered in 2016. Its relative atomic mass is 294. Suggest a reason why its predicted density is 5000 g/dm³. [1]

08 This question is about Group 1 reactions.

08.1 Write a balanced equation for the reaction of sodium with water. Include state symbols. [3]

08.2 Explain why potassium is more reactive than lithium. [3]

09 A student carries out an investigation into the reactions of halogens with their salts. The salts solutions are colourless. The table show the student's results.

	Sodium chloride	Potassium bromide	Potassium iodide
Chlorine	Not done	Changes to orange	Changes to brown
Bromine	No visible change	Not done	Changes to brown
Iodine	No visible change	No visible change	Not done

09.1 Give **one** reason why the student did not investigate three of the possible mixtures. [1]

09.2 Explain what the student's results show about the reactivity of the halogens. [3]

09.3 Write an ionic equation for the reaction between bromine and potassium iodide. [2]

Chemistry only

10 Zinc and copper are placed between groups 2 and 3 in the periodic table. Zinc only forms Zn^{2+} ions but copper forms Cu^+ and Cu^{2+} ions. The table compares some other properties.

	Melting point in °C	Density in g/cm³	Metal oxide as a catalyst	Colour of hydroxide	Colour of chloride
Copper	1085	8.96	Catalyst	Blue	Blue-green
Zinc	693	7.14	Catalyst	White	White

Chemists do not classify zinc as a transition metal, even though it is immediately to the right of copper in the periodic table. Use your knowledge of the typical properties of transition metals, and data in the table, to evaluate the classification of zinc. [6]

CHEMICAL BONDS

There are three types of strong **chemical bonds**: ionic, covalent and metallic.

Metals and non-metals

You can predict the type of bonding in an element or compound if you know whether it contains metals, non-metals or both:

- **Ionic bonding** nearly always involves a compound of a metal and a non-metal.
- **Covalent bonding** exists in most non-metal elements and compounds of non-metals.
- **Metallic bonding** exists in metal elements and their alloys.

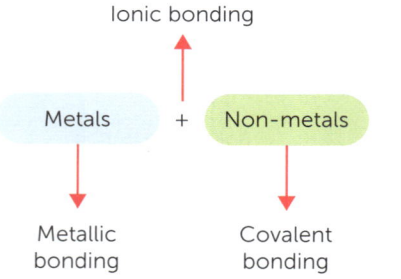

Electrostatic forces of attraction

All chemical bonds involve **electrostatic** forces of attraction between opposite charges. These forces arise in different ways depending on the type of bonding.

Type of bond	Forces of attraction between:
Ionic	Oppositely charged ions
Covalent	Nuclei of two atoms and a shared pair of electrons
Metallic	Nuclei of atoms and delocalised electrons

You can revise these three types of bond in detail: ionic bonding on **page 27**, covalent bonding on **page 29**, and metallic bonding on **page 31**.

---

1. (a) Predict the type of bonding present in sodium chloride. [1]
   (b) Explain why ammonium nitrate has ionic bonding, even though it does not contain metals. [2]
2. Predict the type of bonding present in sulfur dioxide, SO_2. Give a reason for your answer. [2]

   1. (a) Ionic bonding.[1]
      (b) Ammonium nitrate contains NH_4^+ ions[1] and NO_3^- ions[1].
   2. Covalent bonding[1] because it is a compound of two non-metal elements[1].

Transferring and sharing electrons

Chemical bonds form when electrons are transferred or shared.

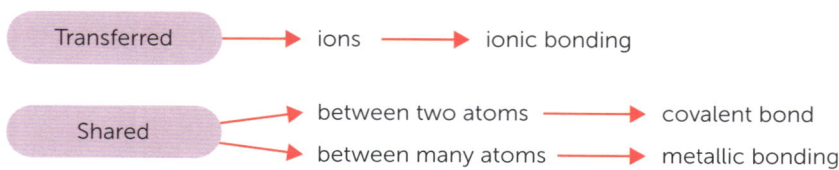

4.2.1.2 | 5.2.1.2

IONIC BONDING

Ions involved in ionic bonding can be represented by **dot and cross diagrams**.

Predicting the charges on ions

In a reaction between a metal and a non-metal, the outer electrons from metal atoms are transferred to the outer shells of non-metal atoms. Metals form positively charged ions and non-metals form negatively charged ions.

The charge on ions formed by atoms in groups 1, 2, 6 and 7 is related to the group number.

	Metals		Non-metals	
Group number	1	2	6	7
Charge on ions	+1	+2	−2	−1

1. Predict the charge on the ions formed by:
   (a) sodium
   (b) calcium
   (c) oxygen
   (d) chlorine [4]

   1. (a) +1[1]
      (b) +2[1]
      (c) −2[1]
      (d) −1[1]

Dot and cross diagrams

Dot and cross diagrams represent the electrons in different atoms and ions. The diagrams show the electronic structures for sodium atoms and sodium ions. Each dot represents an electron. It is usual to use shortened electronic structures in dot and cross diagrams for ionic bonding.

Dots represent electrons in metal atoms and metal ions. Crosses represent electrons in non-metal atoms and non-metal ions (except for electrons that have come from metal atoms).

Full electronic structures

Shortened electronic structures

2. Draw a dot and cross diagram to represent the formation of sodium chloride. [3]
3. (a) Write the electronic structures of (i) a sodium ion, and (ii) a chloride ion. [2]
   (b) Suggest what your answers to part (a) show about these ions. [3]

   2. Correct structures of Na and Cl[1], correct structures of ions[1], correct charges[1].

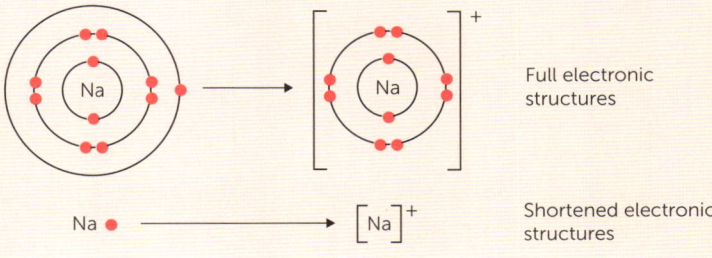

   3. (a) (i) 2,8[1]   (ii) 2,8,8[1]
      (b) They have complete outer shells which gives them the same electronic structures as Group 0 elements (noble gases)[1]. Na^+ has the same electronic structure as neon[1], and Cl^- has the same electronic structure as argon[1].

AQA GCSE **Chemistry** 8462 / 8464 – **Topic 2**

4.2.1.3 | 5.2.1.3

IONIC COMPOUNDS

The ions in **ionic compounds** are held together by **ionic bonding**.

Giant ionic lattice

Ionic compounds have a **giant ionic lattice** structure:
- Lattice – A regular structure.
- Ionic – The structure consists of ions with ionic bonding.
- Giant – The regular structure is repeated very many times.

Ionic bonding acts in all directions in the lattice. It is the strong electrostatic force of attraction between oppositely charged ions.

⭐ You should know the structure of sodium chloride, but not the structures of other ionic compounds

Representing ionic structures

Ionic structures extend in three dimensions. You can represent these structures using plastic molecular modelling kits. Each ball represents an ion and each stick represents the bonding.

You can also show ionic structures in two dimensions.

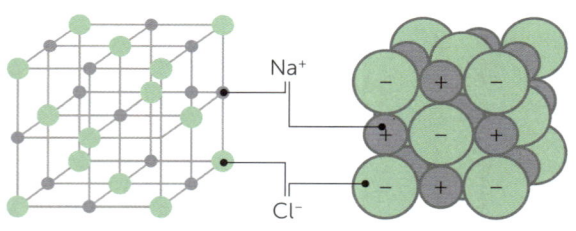

Ball and stick — Space-filling

A third representation of bonding in ionic compounds is through dot and cross diagrams. (See **page 27**.)

The different diagrams have limitations.

	Dot and cross	Ball and stick	Space-filling
Shows bonding	In detail	As lines	✗
Shows relative sizes of atoms	✗	Inaccurately	✓
Shows shape of lattice	✗	✓	✓

The diagram shows the structure of silver bromide.
  (a) Explain why the diagram shows that silver bromide is an ionic compound. [2]
  (b) Define the term **empirical formula**. [2]
  (c) Determine the empirical formula of silver bromide. [2]

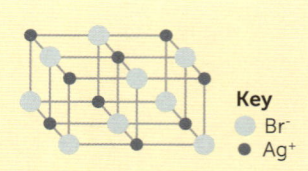

Key
○ Br⁻
● Ag⁺

  (a) It shows oppositely charged ions[1] from more than one element[1].
  (b) The simplest whole number ratio[1] of ions or atoms in a substance[1].
  (c) The diagram shows 9 Ag⁺ ions and 9 Br⁻ ions[1], the simplest whole number ratio is 1 : 1 so the empirical formula is AgBr[1].

4.2.1.4 5.2.1.4

COVALENT BONDING

Molecules of non-metal atoms are held together by **covalent bonds** created by sharing pairs of electrons.

Types of molecules

Substances with covalent bonds exist as molecules. There are three types of molecule:
- Small molecules
- Very large molecules
- Giant covalent structures

A covalent bond is a shared pair of electrons.

Representing molecules

You can represent molecules in two dimensional diagrams. A **displayed structural formula** is a common type of these diagrams. Each atom is shown by its chemical symbol, and each covalent bond is shown as a single line.

Methane (Simple / small molecule)

Poly(ethene) (Very large molecule)

Diamond (Giant structure)

Polymers such as poly(ethene) have a repeating structure. Their molecules consist of very many atoms of two or more elements. You could not sensibly draw every bond and atom label in these molecules. Instead, you show the **repeating unit** in brackets. The letter n stands for a very large number, which could be many hundreds or thousands.

Methane molecules and most other molecules extend in three dimensions. You can represent them using plastic molecular modelling kits. Each ball represents an atom and each stick represents a covalent bond.

These models are often drawn as ball and stick diagrams.

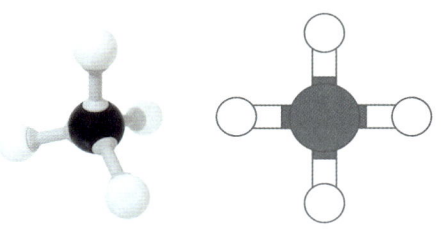

This diagram shows the full displayed structural formula of ethane.

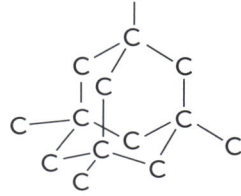

(a) Define the term **molecular formula**. [1]

(b) Determine the molecular formula of ethane. [1]

(a) The actual number of atoms of each element in a molecule, shown as chemical symbols and numbers.[1]

(b) C_2H_6 [1]

AQA GCSE Chemistry 8462 / 8464 – Topic 2

4.2.1.4 5.2.1.4

DOT AND CROSS DIAGRAMS

Dot and cross diagrams can represent the bonding in small molecules.

Representing covalent bonds

The diagram is a dot and cross diagram for a chlorine molecule, Cl_2. In this type of diagram:
- only the outer shell of each atom is shown
- dots or crosses represent electrons in each atom
- a dot and cross represents a single covalent bond
- lone pairs of electrons are the ones that are not in bonds.

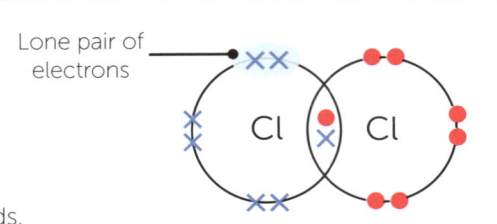

Lone pair of electrons

Drawing a dot and cross diagram

Hydrogen atoms can only form one covalent bond each. For other non-metal elements:

Group	4	5	6	7
Number of bonds	4	3	2	1
Example	C	N	O	Cl

This helps you work out how to complete a dot and cross diagram when you know the chemical formula for a substance. For example, the chemical formula for ammonia is NH_3.

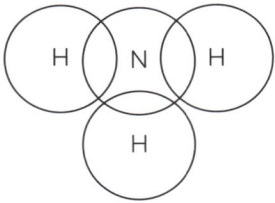 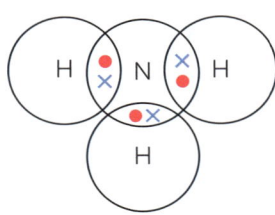 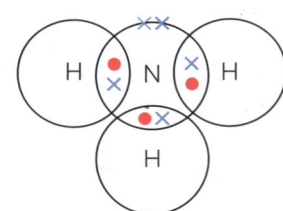

Draw overlapping circles, one for each atom's outer shell | Add a dot and cross for each covalent bond | Make sure that all outer shells are full

---

1. The structural formula of oxygen is O=O.
   (a) Describe what the double line in the formula represents. [1]
   (b) Draw a dot and cross diagram for oxygen. [2]
2. Draw a dot and cross diagram for nitrogen, N_2. [2]

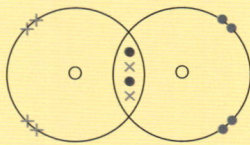 

1. (a) A double covalent bond.[1]
   (b) Two shared pairs[1] and correct outer electrons[1].

2. Three shared pairs[1] and correct outer electrons[1].

METALLIC BONDING

Metals and alloys are held together by **metallic bonding**.

Giant structures

In solid metals, the metal atoms are arranged in regular patterns. Two-dimensional diagrams often show just one layer of atoms. However, there will be many layers of atoms which form a **giant structure** in three dimensions.

Graphite is a non-metal that contains delocalised electrons. You can revise the structure and properties of this form of carbon on **page 38**.

Delocalised electrons

An atom consists of a central nucleus surrounded by **electrons** arranged in energy levels or **shells**. Electrons usually remain with individual atoms. However, outer electrons can become **delocalised** in some chemical structures – they leave individual atoms and become free to move through some or all of the structure.

Positively charged **ions** form when electrons permanently leave atoms. However, delocalised electrons do not permanently leave atoms. They are just free to move from atom to atom instead. Overall, the number of electrons in each atom remains equal to the number of **protons**. The outer electrons in metals are delocalised, and this gives rise to metallic bonding.

Representing metallic bonding

Metallic bonding is the electrostatic force of attraction between the positively charged centres of metal atoms and a 'sea' of delocalised electrons.

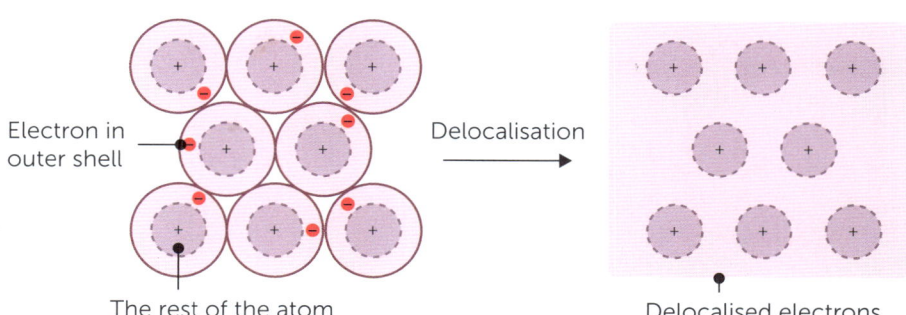

---

1. Give the location of the electrons that become delocalised in metals. [1]
2. Compare metallic bonding with ionic bonding. [3]

   1. The outer shells.[1]
   2. Both involve electrostatic forces of attraction[1] and both act in all directions[1]. However, ionic bonding involves oppositely charged ions but metallic bonding involves atoms and delocalised electrons[1].

THE THREE STATES OF MATTER

4.2.2.1 | 5.2.2.1

The three **states of matter** are **solid**, **liquid** and **gas**.

The particle model

The **particle model** represents the arrangement and movement of particles in the three different states. The particles can be atoms, ions or molecules, but all of them are shown as small, solid spheres.

	Solid	Liquid	Gas
Arrangement of particles	Regular	Random	Random
Relative distance between particles	Close	Close	Far apart
Motion of particles	Vibrate about fixed positions	Move around each other	Move quickly in all directions
Particle diagram			

Energy and forces

For a given sample of a substance:

Solid → Liquid → Gas
Increasing energy

When energy is transferred to a substance by heating, particles gain energy and the temperature of the substance increases. When enough energy is transferred, forces between particles are overcome and the substance changes state:
- Melting occurs when some of the forces are overcome.
- Boiling occurs when all of the forces are overcome.

The stronger the forces are between particles, the higher the **melting point** and **boiling point** of the substance. This is because more energy is needed to overcome strong forces than weak forces.

1. Name the state changes that happen at:
   (a) The melting point. [1]
   (b) The boiling point. [1]
2. Compare the arrangement and movement of particles in:
   (a) The solid state and the liquid state. [2]
   (b) The liquid state and the gas state. [2]

1. (a) Melting and freezing.[1]
   (b) Boiling and condensing.[1]
2. (a) Particles in both states are close together[1] but they are regularly arranged in the solid state and randomly arranged in the liquid state[1].
   (b) Particles in both states are randomly arranged[1] but they are close together in the liquid state and far apart in the gas state[1].

USING THE PARTICLE MODEL

The particle model explains the temperatures at which state changes happen.

Predicting states

You can predict the state of a substance at a given temperature if you know its melting and boiling points:
- Solid below the melting point
- Liquid between the melting and boiling points
- Gas above the boiling point

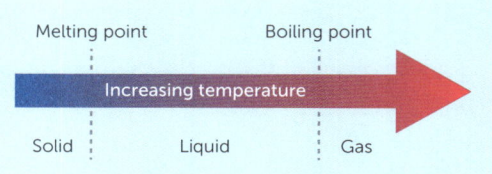

State change and temperature

The temperatures at which state changes happen depend on the type of structure and bonding present in a substance. Weak forces between small molecules are overcome during melting and boiling.

Type of substance	Forces overcome	Strength of forces	Melting and boiling points
Small molecule	Intermolecular	Weak	Low
Large molecule	Intermolecular	Medium to strong	Medium to high
Giant ionic	Ionic	Strong	High
Giant covalent	Covalent	Strong	High
Metal	Metallic	Strong	High

Bulk properties

A **bulk property** of a substance is one of its observable or measurable features. Melting point and boiling point are examples of bulk properties. Properties like these can be explained by considering the particles in a substance, but the particles themselves do not have bulk properties.

Limitations of the particle model — Higher Tier only

The simple particle model describes all particles as solid, inelastic spheres. However, particles can change shape slightly when they are close together or rebound off each other. Simple ions and individual atoms are spherical, but most ions and molecules have complex three-dimensional shapes.

---

1. The melting point of ethene is −169 °C. The melting point of poly(ethene) is around 120 °C. Explain how the molecular structure of ethene and poly(ethene) influences their melting point. [4]
2. **Higher Tier only:** Describe **one** of the limitations of the simple particle model. [2]

   1. Ethene exists as small molecules[1] with few, weak forces between them[1]. However, poly(ethene) exists as large molecules[1] with many more weak forces between them[1].
   2. It assumes that no forces exist between particles in the gas state[1] but very weak forces still exist[1].

PROPERTIES OF IONIC COMPOUNDS

Melting point and boiling point

Ionic compounds have some typical bulk properties with high melting points and boiling points. For example, aluminium oxide (alumina) melts at 2072 °C and boils at 2977 °C. These temperatures are so high that alumina bricks are used to line the steel walls of blast furnaces (see **page 61** on the extraction of iron).

Ionic compounds have a giant ionic lattice structure. There are electrostatic forces of attraction between oppositely charged **ions**. This ionic bonding is strong, and it acts in all directions.

Large amounts of energy must be transferred to ionic compounds in order to overcome the strong electrostatic forces of attraction between the ions. This is why ionic compounds have high melting and boiling points.

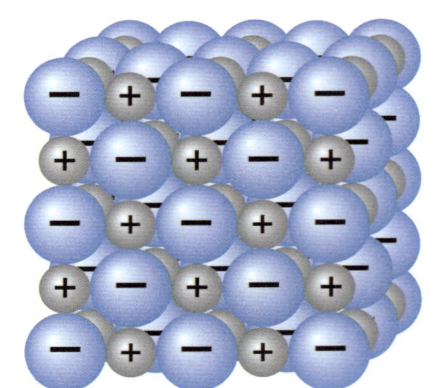

Conducting electricity

A substance can conduct **electricity** if:
- it contains charged particles, and
- these particles are free to move from place to place.

Ionic compounds do contain charged particles – their ions. However, these particles are not always able to move from place to place. They are only free to move from place when the substance is in the liquid state, but not when it is in the solid state.

1. Explain why ionic compounds are able to conduct electricity in aqueous solution. [1]
2. Copper(II) oxide is insoluble in water. Give **one** way to allow copper(II) oxide to conduct electricity. [1]

   1. Their ions are separate and free to move[1] when dissolved in water, so charge can flow[1].
   2. Melt it.[1]

Ions move to oppositely charged plates in liquids

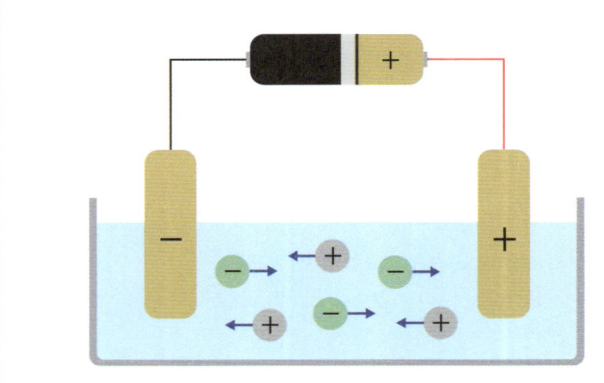

Electricity is conducted by ionic compounds during electrolysis. In this process, ions move to oppositely charged electrodes. Once there, they lose or gain electrons and become discharged as atoms or molecules. You can revise electrolysis on **pages 73-76**.

PROPERTIES OF SMALL MOLECULES

Small molecules

The size of molecules varies between substances. This affects the physical properties of substances. Substances that consist of small molecules have relatively low melting and boiling points. These substances may be elements or compounds, and their molecules contain few atoms.

The atoms in a small molecule are held together by strong covalent bonds. These bonds are not broken during melting or boiling. Instead, the much weaker intermolecular forces between molecules are overcome. In general, as the size of the molecule increases, intermolecular forces increase and therefore, melting and boiling points increase.

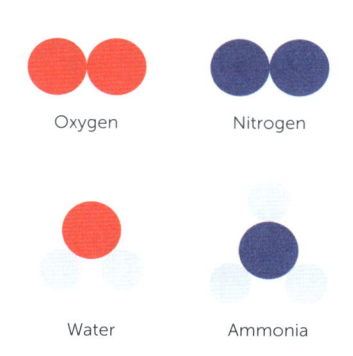

Substance	CH_4	C_2H_6	C_3H_8	C_4H_{10}
Boiling point in °C	−161	−89	−42	−1

POLYMERS

Chemistry only: Addition polymerisation is covered on **page 117**.

Chemistry Higher only: Condensation polymerisation is covered on **page 118**.

Polymers consist of large molecules, formed from many small molecules called **monomers**. The atoms in each individual polymer molecule are joined together by strong covalent bonds. Polymer molecules are attracted to each other by intermolecular forces. These forces are relatively strong because polymer molecules are so large.

You should be able to recognise polymers from diagrams like these.

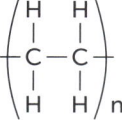
Repeating unit

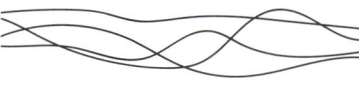
Tangled polymer chains

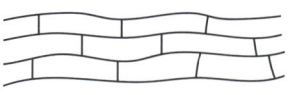
Chains with cross-links

Explain why poly(propene) is in the solid state at room temperature. [4]

Poly(propene) consists of large molecules[1]. Intermolecular forces between polymer molecules are overcome during melting[1]. These are relatively strong because the molecules are so large[1] so a lot of energy is needed to overcome them[1].

AQA GCSE **Chemistry** 8462 / 8464 – Topic 2

4.2.2.6 5.2.2.6

GIANT COVALENT STRUCTURES

Some elements and compounds have **giant covalent structures**.

Structure and bonding

Giant covalent structures have a lattice structure in which:
- all the atoms are linked to other atoms by covalent bonds.
- the regular lattice structure is repeated very many times.

Covalent bonds are strong and there are many of them in a giant covalent structure. A lot of energy must be transferred to melt or boil a substance with this structure. As a result of this, substances with giant covalent structures have high melting points and boiling points. For example, **sand** is mostly silica, a compound with a giant covalent structure. This melts at 1713 °C and boils at 2950 °C.

Recognising giant covalent structures

Diamond and graphite are two forms of carbon. They both have giant covalent structures, although these are different from each other (see **page 38**).

You should be able to recognise giant covalent structures from diagrams. There are very many atoms in a giant covalent structure, so you will see just enough atoms for you to understand how they are arranged. Remember that the structure you see will be repeated many times, with covalent bonds leading from atoms at the edges of the diagrams to atoms that are not shown.

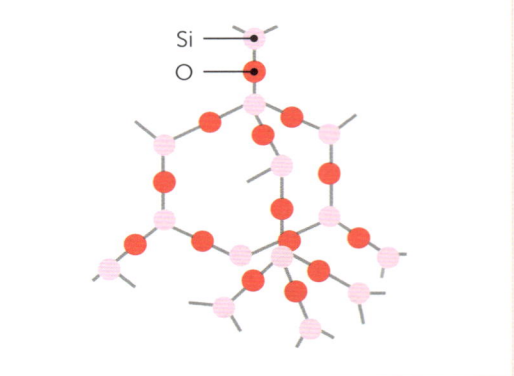

The structure of silica, SiO_2

The diagram shows the structure of a form of boron nitride, BN.

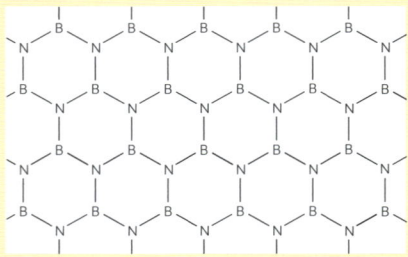

(a) Give the type of structure of boron nitride. Explain your answer. [3]
(b) Give **one** physical property that you would expect boron nitride to have. [1]

(a) Giant covalent structure[1] because it contains many atoms[1] joined by covalent bonds[1].
(b) One from: high melting point[1], high boiling point[1].

PROPERTIES OF METALS AND ALLOYS

Bending and shaping metals

Metals are good conductors of electricity and **thermal energy**. They are:
- **Malleable** – they can be bent or hammered into shape without breaking
- **Ductile** – they can be pulled to make wires without snapping.

Pure metals have these properties because their layers of atoms can slide over each other when a force is applied.

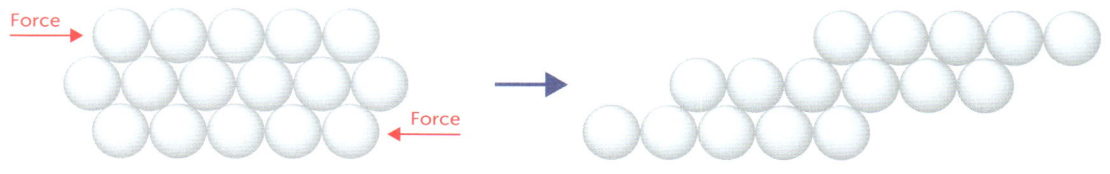

Hardness

A pure metal element may be too soft for a lot of uses. An **alloy** is a **mixture** of a metal element with at least one other element, usually another metal. Some alloys are harder and therefore more useful than the pure metal.

For example, steels are mixtures of iron with carbon and some other elements. Steels are much harder than iron. This is because the carbon atoms distort the regular lattice structure, making it more difficult for the layers to slide over one another.

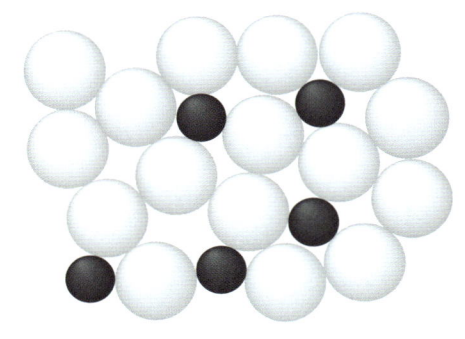

Conducting electricity

A substance can conduct **electricity** if it contains charged particles, and if these particles are free to move from place to place.

Metals contain **delocalised electrons**. These charged particles can move freely though the giant structure of metal atoms. This is why metals are good conductors of electricity.

1. Explain why metals are good conductors of thermal energy. [3]
2. Explain, in terms of structure and bonding, why metals have high melting points. [3]

> 1. When a metal is heated, energy is transferred to delocalised electrons[1]. These are free to move through the structure of the metal[1] transferring energy to atoms and other electrons[1].
>
> 2. Metals in the solid state have giant structures of atoms[1] held together by metallic bonding[1]. A lot of energy must be transferred to overcome this strong bonding[1].

DIAMOND AND GRAPHITE

Diamond and **graphite** are different forms of carbon with different giant covalent structures.

Structures

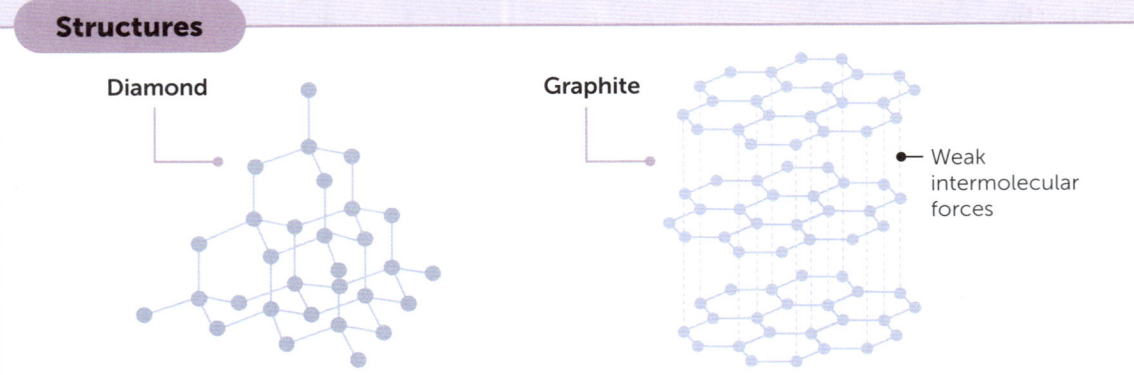

Property	Diamond	Graphite
Contains covalent bonds	✓	✓
Atoms bonded to each C atom	4	3
Contains layers	✗	✓ Hexagonal rings of atoms
Weak forces between layers	✗	✓
Contains delocalised electrons	✗	✓

Diamond

Diamond has a very high melting point because its structure contains very many strong covalent bonds. A lot of energy is needed to break these bonds. Diamond does not contain delocalised electrons or any other charged particles that are free to move, so it does not conduct electricity.

Graphite

There are only weak intermolecular forces, rather than strong covalent bonds, between the layers of atoms in graphite. These allow the layers to slide over each other easily, so graphite can be slippery.

---

1. Explain why diamond is very hard. [3]
2. (a) Explain, in terms of particles, why graphite is similar to metals. [2]
   (b) Explain why graphite is a good conductor of electricity. [2]

   1. Diamond has a giant covalent structure[1]. Its many strong covalent bonds[1] resist forces that could distort the structure[1].
   2. (a) Both consist of atoms[1] and contain delocalised electrons[1].
      (b) One electron from each carbon atom becomes delocalised.[1] These delocalised electrons are charged and are free to move.[1]

GRAPHENE AND FULLERENES

Graphene and fullerenes are different forms of carbon.

Graphene

Graphene is a single layer of graphite. It does not have a layered structure, so it is transparent and flexible.

Graphene does have delocalised electrons so it conducts electricity, just like graphite does. This property is useful in electronics.

The carbon atoms in graphene are joined by strong covalent bonds, so graphene itself is strong. This property makes it useful as a replacement for carbon fibres in composite materials. You can revise these materials on **page 157**.

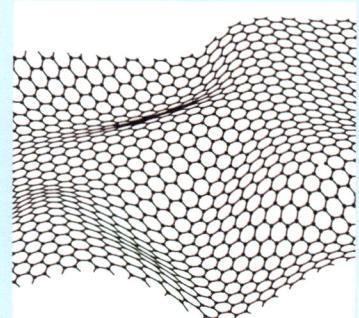

Fullerenes

Fullerenes are hollow molecules of carbon. **Buckminsterfullerene**, C_{60}, was the first fullerene to be discovered. Its molecules consist of 60 carbon atoms arranged in rings of five or six atoms. Fullerenes may also contain rings of seven carbon atoms.

Carbon **nanotubes** are cylindrical fullerenes. They have very high length to diameter ratios – this means that they are very long compared to their width.

A section of a cylindrical fullerene	A Buckminsterfullerene molecule
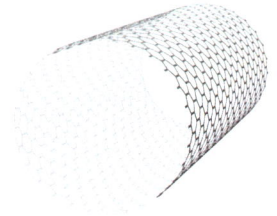	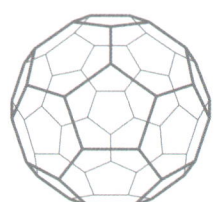

1. (a) Explain why fullerenes can conduct electricity. [2]
   (b) Give **one** potential use of fullerenes that depends on their ability to conduct electricity. [1]
2. Explain why a mixture of carbon nanotubes and plastic are used in wind turbine blades rather than plastic alone. [3]
3. Suggest **one** reason why Buckminsterfullerene may be used in lubricants. [1]

1. (a) They contain delocalised electrons[1] which can move through their structures[1].
   (b) Electronics.[1]
2. Carbon nanotubes are very strong[1] and stiff[1]. They can be used to reinforce the plastic[1] as a composite material[1] instead of carbon fibres. Composite materials are strong and flexible[1].
3. Its molecules are spherical and can roll around each other.[1]

4.2.4.1 Chemistry

NANOPARTICLES

Sizes of particles

Nanoscience involves particles that contain just a few hundred atoms. Small particles can be placed into different categories according to their size.

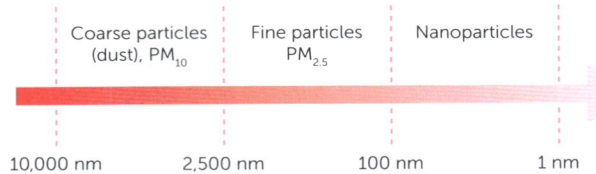

The sizes of these particles differ by several **orders of magnitude**. For example, a PM_{10} (Particulate Matter) particle has a diameter of 9000 nm and a nanoparticle has a diameter of 60 nm:

$$\frac{9000 \text{ nm}}{60 \text{ nm}} = 150$$

$150 = 1.5 \times 10^2$, so the two particles differ in size by about two orders of magnitude.

Calculating surface area to volume ratios

Nanoparticles have very large **surface area to volume ratios**. For example, the side length of a cube-shaped silver nanoparticle is 10 nm.

Surface area = 6 sides × 10 nm × 10 nm = 600 nm²

Volume = 10 nm × 10 nm × 10 nm = 1000 nm³

Surface area to volume ratio = $\frac{600}{1000}$ = 0.6

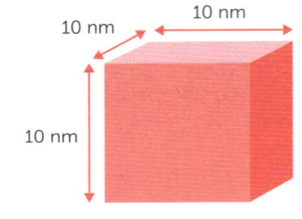

The surface area to volume ratio of a cube is inversely proportional to its side length:

As its side length goes down by a factor of 10, its surface area to volume ratio goes up by a factor of 10.

This means that the surface area of a given mass of nanoparticles is vast compared to the same mass of the substance in bulk. As a result of this, smaller quantities of **nanoparticulate** materials may be needed for the same purpose as lumps or sheets of the same material.

---

1. Explain why the properties of nanoparticles may differ from the properties of the same substance in bulk. [2]
2. Convert the following quantities into metres, expressed in standard form:
   (a) 10,000 nm [1]   (b) 2500 nm [1]   (c) 100 nm [1]

1. Nanoparticles are only 1–100 nm in size[1] so their surface area to volume ratios are high[1].
2. (a) 1×10^{-5} m.[1]   (b) 2.5×10^{-6} m.[1]   (c) 1×10^{-7} m.[1]

4.2.4.2 Chemistry

USES OF NANOPARTICLES

Sizes of nanoparticles

Scientists are researching practical applications of **nanoparticulate** materials. Nanoparticles are larger than atoms and small molecules, but smaller than red blood cells and bacteria. They are similar in size to viruses and to large molecules such as proteins.

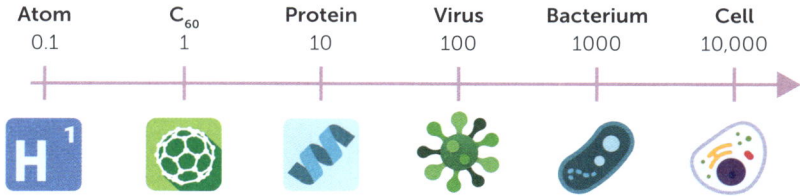

Approximate size in nm

The use of nanoparticles has possible risks. They can enter the body, just as fine and coarse particles of bulk materials can. This may be through cuts, and when you eat or breathe. Their small sizes make it possible that they could enter cells. They might carry toxic substances into the cells or catalyse damaging reactions. Scientists are not sure how high these risks are.

Uses of nanoparticles

The applications of nanoparticles depend upon their physical and chemical properties, which may differ from the same substance in bulk. The table show some typical examples.

Application	Example	Small size	High surface area to volume ratio	High length to diameter ratio
Electronics	Nanorods in transistors			✓
Cosmetics	Almost transparent sunscreens	✓		
Deodorants	Antibacterial sports socks		✓	
Medicines	Targeting cancer cells for destruction	✓		✓
Catalysts	Self-cleaning windows		✓	

1. Titanium dioxide absorbs harmful ultraviolet light. It is white in bulk but almost transparent as nanoparticles. Suggest **one** advantage and **one** disadvantage of using nanoparticles in sunscreens. [2]
2. Silver has antibacterial properties. Suggest a reason that explains why silver nanoparticles may be used instead of bulk silver. [3]

   1. Advantage: sunscreen is less visible on the skin.[1] Disadvantage: it is difficult to see where you have applied it.[1]
   2. Silver is expensive[1] but silver nanoparticles have high surface area to volume ratios[1] so smaller quantities of silver are needed[1].

TOPIC 2

EXAMINATION PRACTICE

01 One of the stages in making steel involves adding magnesium powder. This reacts with sulfur impurities to form an ionic compound called magnesium sulfide.
The diagram shows the outer electrons in magnesium and sulfur atoms.

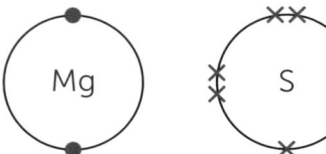

01.1 Predict the formula of magnesium sulfide. [1]
01.2 Describe, in terms of electron transfer, what happens when magnesium reacts with sulfur. [3]
01.3 Describe the bonding in magnesium sulfide. [2]
01.4 Predict **two** physical properties of magnesium sulfide. [2]

02 Methane, CH_4, is the main component of natural gas.
02.1 Complete the dot and cross diagram to show the bonding in a methane molecule. [2]

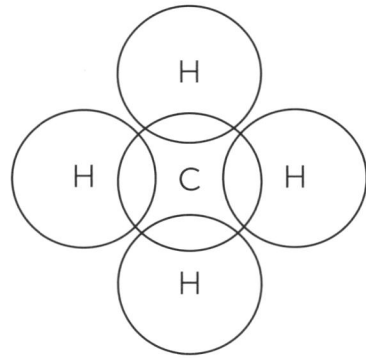

02.2 Explain why methane is in the gas state at room temperature. [3]

03 Diamond, graphite and graphene are three forms of carbon.
03.1 Compare the structure and bonding in diamond and graphite. [4]
03.2 Explain why graphite and graphene can conduct electricity. [2]

04 This question is about aluminium and its alloys.
04.1 Explain why most metals have high melting points. [2]
04.2 Explain why pure aluminium can be bent and shaped without breaking. [2]
04.3 Aluminium-lithium alloys consist of aluminium and around 2.5% lithium.
Explain why these alloys are stronger and harder than pure aluminium. [2]
04.4 Aluminium is used to make high-voltage overhead power lines. It is also used to make cooling units in computers.
Explain why aluminium is a good conductor of thermal energy. [2]

05 The diagram shows the crystal structure of wüstite, a mineral form of an oxide of iron.

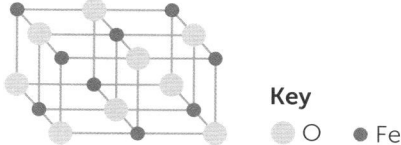

Key
○ O ● Fe

05.1 Describe what is meant by an empirical formula. [2]

05.2 Determine the empirical formula of wüstite. [1]

06 The diagram shows the structure of a compound.

$$\left(\begin{array}{c}F \;\; H\\ |\;\;\;\;|\\ -C-C-\\ |\;\;\;\;|\\ F\;\; H\end{array}\right)_n$$

06.1 Name the type of substance shown in the diagram. [1]

06.2 Give the meaning of **one** of the straight lines in the diagram. [1]

06.3 Explain why compounds like this one are solid at room temperature. [2]

07 Dodecanoic acid is in the solid state at room temperature. A student heated a sample of this substance and recorded its temperature at regular intervals. The graph shows the results.

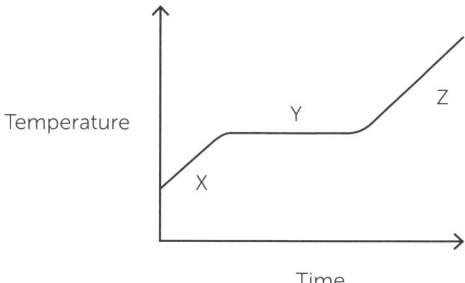

Explain the results in terms of the arrangement, energy and movement of dodecanoic acid particles. [6]

Chemistry only

08 A type of self-cleaning window glass is coated with a layer of cube-shaped vanadium oxide nanoparticles. These help water run off the glass and catalyse the breakdown of dirt.

08.1 Explain what is meant by a nanoparticle. [2]

08.2 Describe a property of nanoparticles that make them useful as catalysts. [1]

08.3 The mean side length of the vanadium oxide nanoparticles is 25 nm.
Describe how to increase their surface area to volume ratio by a factor of 10. [2]

RELATIVE FORMULA MASS

The **law of conservation of mass** states that the total mass of the products formed in a reaction is the same as the total mass of the reactants used.

Calculating relative formula mass, M_r

Covalent compounds and ionic compounds have a **relative formula mass**, and so do elements that exist as molecules. You can calculate a relative formula mass by following these steps:
- look at the **relative atomic masses** of all the elements in the formula
- add together the relative atomic masses for all the atoms in the formula.

Just like relative atomic masses, relative formula masses are numbers without units.

Law of conservation of mass

Mass is conserved during a chemical reaction because no atoms are lost or made in the reaction. This is why you need to balance symbol equations. In a balanced equation the numbers of atoms of each element must be the same on both sides of the arrow.

For example, methane burns completely in air to produce carbon dioxide and water:

$$CH_4 + 2O_2 \rightarrow CO_2 + 2H_2O$$

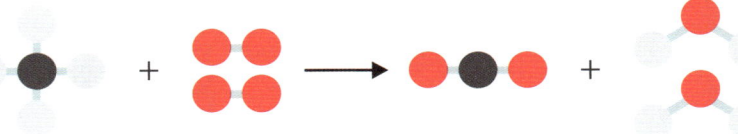

M_r and balanced equations

When a chemical equation is balanced, the total M_r of all the reactant particles equals the total M_r of all the product particles. In the example above:

Relative formula masses: $CH_4 = 16$, $O_2 = 32$, $CO_2 = 44$, $H_2O = 18$

Total M_r of reactants $= 16 + (2 \times 32) = 80$ **Total M_r of products** $= 44 + (2 \times 18) = 80$

---

1. Calculate the relative formula mass (M_r) of ammonium carbonate $(NH_4)_2CO_3$.
   Relative atomic masses (A_r): hydrogen = 1, carbon = 12, nitrogen = 14, oxygen = 16. [2]
2. In a chemical reaction, 6.0 g of carbon burns in excess oxygen to produce 22 g of carbon dioxide.
   Determine the mass of oxygen used in the reaction. [1]
3. Balance this equation: $Al(s) + O_2(g) \rightarrow Al_2O_3(s)$ [2]

   1. 96[1] Evidence of working[1], e.g. $(2 \times 14) + (2 \times 4 \times 1) + 12 + (3 \times 16)$
   2. $22 - 6.0 = 16$ g[1]
   3. $4Al(s) + 3O_2(g)$[1] $\rightarrow 2Al_2O_3(s)$.[1]

PERCENTAGE COMPOSITION BY MASS

The percentages by mass of each element in a compound add up to 100%. To calculate a **percentage composition** by mass, you need to know the relative atomic mass of the element you are interested in and the relative formula mass of the compound.

Calculating percentage composition

Methane, CH_4, is a compound of carbon and hydrogen. Its molecules each contain one carbon atom and four hydrogen atoms. You might think that the percentage of carbon is 20% and the percentage of hydrogen is 80%. However, you must take into account the relative atomic masses of each element.

Percentage by mass of hydrogen = $\dfrac{\text{(number of atoms of H)} \times (A_r \text{ of H})}{M_r \text{ of } CH_4} \times 100$

= $\dfrac{4 \times 1}{12 + (4 \times 1)} \times 100 = \dfrac{4}{16} \times 100 = 25\%$

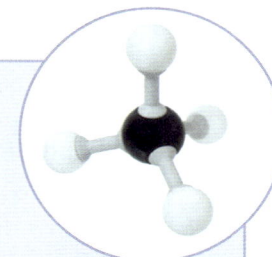

A more difficult example

Ammonium nitrate, NH_4NO_3, is widely used in blast explosives and as a high-nitrogen fertiliser in agriculture. Calculate its percentage by mass of nitrogen.

Relative atomic mass (A_r) of nitrogen = 14, relative formula mass (M_r) of ammonium nitrate = 80.

Percentage by mass of nitrogen = $\dfrac{2 \times 14}{80} \times 100 = 35\%$.

1. The percentage by mass of carbon in ethane C_2H_6 is 80%. Determine the percentage by mass of hydrogen in ethane. [1]
2. Calculate the percentage by mass of oxygen in sulfur dioxide, SO_2. Relative atomic mass (A_r) of oxygen = 16, relative formula mass (M_r) of SO_2 = 64. [2]
3. Calculate the percentage by mass of hydrogen in copper(II) hydroxide, $Cu(OH)_2$. Relative atomic mass (A_r) of hydrogen = 1.00, relative formula mass (M_r) of $Cu(OH)_2$ = 97.5. [2]

   1. 100 − 80 = 20%.[1]
   2. 50%[1] Evidence of working[1], e.g. % of O = $\dfrac{(2 \times 16)}{64} \times 100$
   3. 2.05%[1] Evidence of working[1], e.g. % of H = $\dfrac{(2 \times 1.00)}{97.5} \times 100$

⭐ Give your answers to a suitable number of significant figures. This is usually the lowest number of significant figures in the quantities used.

AQA GCSE **Chemistry** 8462 / 8464 – Topic 3

4.3.1.3 5.3.1.3

MASS CHANGES WITH GASES

Reactions involving gases may appear to involve a change in mass. This is usually explained when a reactant or product is a gas which has not been accounted for.

Closed systems

Substances cannot enter or leave a **closed system**. Common closed systems in chemistry include test tubes or flasks with bungs on and **precipitation** reactions.

In a precipitation reaction, an insoluble solid forms when two solutions are mixed. Nothing enters or leaves the reaction mixture, so its total mass stays the same.

In **open systems** (also called **non-enclosed systems**) reactants are free to enter, and products are free to leave. This gives a false impression that the law of conservation of mass has been broken.

Apparent increases in mass

Mass appears to increase when a metal reacts with oxygen from the air to form a metal oxide. This is because oxygen atoms combine with metal atoms in the reaction. If the mass of oxygen gained is included, the law of conservation of mass is not broken.

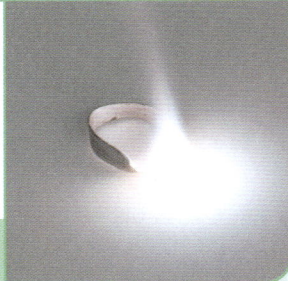

Magnesium burns in air: $2Mg(s) + O_2(g) \rightarrow 2MgO(s)$

You need to be able to explain mass changes in non-enclosed systems, in terms of particles, when given the balanced equation for a reaction. You can revise the particle model on **page 32**.

Apparent decreases in mass

Mass appears to decrease in **thermal decomposition** reactions. In these reactions, a metal carbonate breaks down to form a metal oxide and carbon dioxide, which escapes into the air. If the mass of carbon dioxide lost is included, the law of conservation of mass is not broken.

---

1. Calcium oxide reacts with carbon dioxide in the air: $CaO(s) + CO_2(g) \rightarrow CaCO_3(s)$
   Give **one** reason why the mass of the solid increases. [1]
2. Sodium reacts with water in an open beaker: $2Na(s) + 2H_2O(l) \rightarrow 2NaOH(aq) + H_2(g)$
   Explain why the mass of the reaction mixture goes down. [2]

   1. *Carbon dioxide molecules combine with the calcium oxide.*[1]
   2. *Hydrogen gas is produced in the reaction*[1] *and leaves the reaction mixture*[1].

4.3.1.4 5.3.1.4

CHEMICAL MEASUREMENTS

There is always some **uncertainty** about the result of a measurement.

Resolution

The **resolution** of a measuring instrument is the smallest change that it can show. It is equal to the smallest scale division on the instrument. For example:
- 0.1 g in a digital balance reading to 1 decimal place
- 1 mm in a ruler divided into tenths of a centimetre.

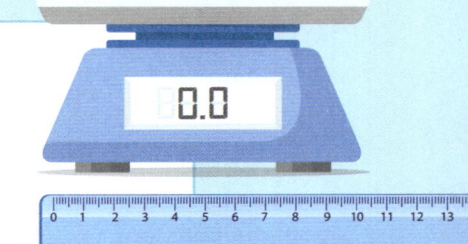

Uncertainty in readings

Uncertainty is given as ± (plus or minus) a value. The uncertainty in readings depends on the type of instrument and its resolution:
- For a digital instrument, the uncertainty is equal to plus or minus half the resolution
- For an instrument with a scale, the uncertainty is equal to the resolution (because you will be taking two readings, the start point and the end point).

1. A digital balance reads 20.4 g. Determine where the true mass lies. [1]
2. A ruler shows that a pencil is 16.8 cm long. Determine where its true length lies. [1]

1. Between 20.35 g and 20.45 g[1]
2. Between 16.7 cm and 16.9 cm[1].

Range and uncertainty in repeats

Uncertainty leads to random errors in readings. You cannot correct for these errors but you can reduce their effects by repeating experiments. **Anomalous** readings can be ignored when calculating a mean value, and this increases the **reliability** of your results.

The **range** of a set of values is the difference between the maximum value and minimum value. The uncertainty in a mean value obtained from a set of repeats is equal to half the range.

3. A student carried out an investigation to determine the volume of carbon dioxide produced in a reaction. The student obtained four repeats. The table shows the results.

Experiment	1	2	3	4
Volume of gas in cm^3	64.1	68.3	66.9	65.5

(a) Calculate the mean volume of carbon dioxide obtained. [1]
(b) Determine the range in volumes. [1]
(c) Give the mean volume, including the uncertainty in this value. [1]

3. (a) "Mean volume = $\frac{(64.1 + 68.3 + 66.9 + 65.5)}{4} = \frac{264.8}{4}$ = 66.2 cm^3[1]
(b) Minimum value = 64.1 cm^3. Maximum value = 68.3 cm^3.
Range = (68.3 − 64.1) = 4.2 cm^3[1]
(c) 66.2 ± 2.1 cm^3[1]

AQA GCSE Chemistry 8462 / 8464 – Topic 3

4.3.2.1　5.3.2.1　**Higher Tier**

MOLES

The **mole** is the unit for an amount of substance. The unit symbol is mol.

Amount of substance

In everyday use, the word 'amount' can apply to quantities such as mass or volume, but it has a very particular meaning in chemistry. The **amount** of a substance refers to the number of particles it contains.

One mole, 1 mol, of any particle contains 6.02×10^{23} of these particles. The particles can be atoms, molecules, ions or electrons. The number of stated particles in 1 mol is called the **Avogadro constant**:

$$\text{Avogadro constant} = 6.02 \times 10^{23} \text{ per mole}$$

Amounts in molecules and compounds

You must be careful to identify the particles involved when you use the mole. For example, a carbon dioxide molecule consists of 1 carbon atom and 2 oxygen atoms. This means that 1 mol of CO_2 molecules contains:
- 1 mol of carbon atoms
- 2 mol of oxygen atoms
- (1 + 2) = 3 mol of atoms.

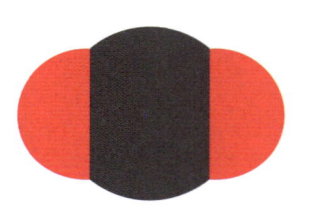

Moles and masses

The mass of 1 mol of a substance is equal to its relative formula mass (M_r) in grams. The M_r of carbon dioxide is 44, so 1 mol of carbon dioxide molecules has a mass of 44 g.

You can use this equation to calculate the mass of a substance in a given amount of it:

$$\text{mass (g)} = M_r \times \text{amount (mol)}$$

You can also use this equation to calculate the amount of a substance in a given mass.

---

1. The relative formula mass (M_r) of calcium hydroxide $Ca(OH)_2$ is 74.
   (a) Calculate the mass of 0.5 mol of calcium hydroxide. [1]
   (b) Determine the amount in moles of OH^- ions in 0.75 mol of calcium hydroxide. [1]
2. Calculate the amount in moles of water molecules in 45 g of water.
   Relative formula mass (M_r) of water = 18 [2]
3. Calculate the number of ammonia molecules in 2 mol of ammonia. [1]

   1. (a) Mass = 74 × 0.5 = 37 g [1]
      (b) Amount = 2 × 0.75 = 1.5 mol [1]
   2. 45 g = 18 × amount (mol), rearranging: amount = $\frac{45}{18}$ [1] = 2.5 mol [1]
   3. Number = amount (mol) × Avogadro constant = 2 × 6.02×10^{23} = 1.204×10^{24} [1]

4.3.2.2 | 5.3.2.2 | **Higher Tier**

AMOUNTS OF SUBSTANCES IN EQUATIONS

Balanced equations can be used to calculate masses of reactants and products.

Interpreting chemical equations

You can interpret balanced symbol equations in terms of moles rather than atoms and molecules. For example, nitrogen reacts with hydrogen to produce ammonia:

$$N_2 + 3H_2 \rightarrow 2NH_3$$

This shows that 1 mol of nitrogen reacts with 3 mol of hydrogen to produce 2 mol of ammonia.

Calculating masses in reactions

You can calculate the mass of a given substance involved in a reaction if you know:
- the balanced symbol equation
- the mass of one of the other substances involved, and
- the relative masses of these two substances.

In the reaction between nitrogen and hydrogen, you could calculate the mass of the nitrogen or hydrogen needed to make a given mass of ammonia, or the mass of ammonia that could be made from a given mass of nitrogen or hydrogen.

An example

Calculate the mass of hydrogen needed to make 13.6 g of ammonia.
Relative formula masses (M_r): H_2 = 2, NH_3 = 17.0

1. Work out the amount in moles of ammonia by rearranging the formula on **page 48**:
$$\text{amount (mol)} = \frac{\text{mass (g)}}{M_r} = \frac{13.6}{17}$$
$$= 0.8 \text{ mol}$$

2. From the balanced equation, 2 mol of NH_3 is made from 3 mol of H_2, so:
$$0.8 \text{ mol of } NH_3 \text{ is made from } \frac{0.8}{2} \times 3 = 1.2 \text{ mol of } H_2$$

3. Work out the mass of hydrogen: mass (g) = M_r × amount (mol) = 2 × 1.2 = 2.4 g

---

Aluminium powder reacts with oxygen to produce aluminium oxide: $4Al + 3O_2 \rightarrow 2Al_2O_3$
Calculate the mass of aluminium oxide that can be made from 8.1 g of aluminium.
Relative masses: Al = 27, Al_2O_3 = 102 [3]

Moles of Al = $\frac{8.1}{27}$ = 0.3 mol[1]
From the balanced equation, mole ratio Al : Al_2O_3 = 4 : 2 so $\frac{0.3}{4} \times 2$ = 0.15 mol of Al_2O_3 [1]
Mass of Al_2O_3 = 102 × 0.15 = 15.3 g[1] Correct answer scores 3 marks[3]

AQA GCSE **Chemistry 8462 / 8464** – Topic 3

4.3.2.3 | 5.3.2.3 | **Higher Tier**

BALANCING EQUATIONS USING MOLES

You can find balancing numbers using masses of reactants and products.

Balancing numbers

Balancing numbers are used in symbol equations to ensure that the numbers of atoms of each element in the reactants and products are the same. You can calculate balancing numbers if you know:
- the masses of the reactants and products
- the relative masses of these substances.

Titanium and its alloys are used to produce strong, corrosion-resistant and lightweight parts such as replacement knee joints. You can revise uses of alloys on **page 154**.

An example

One of the steps in the extraction of the metal titanium involves heating sodium with titanium(IV) chloride. In one of these reactions, 9.2 g of Na reacts with 19 g of $TiCl_4$ to produce 23.4 g of NaCl and 4.8 g of Ti. Determine the balanced symbol equation for this reaction.

Relative masses: Na = 23, $TiCl_4$ = 190, NaCl = 58.5, Ti = 48

You need to follow three steps to answer this question. Remember that:

$$\text{amount (mol)} = \frac{\text{mass (g)}}{\text{relative mass}}$$

Step	Action					
		Na	$TiCl_4$		Na	$TiCl_4$
1	Calculate the amount of each substance	$\frac{9.2}{23}$ = 0.4 mol	$\frac{19}{190}$ = 0.1 mol	→	$\frac{23.4}{58.5}$ = 0.4 mol	$\frac{4.8}{48}$ = 0.1 mol
2	Find the smallest whole number ratio of moles	4	1		4	1
3	Write the equation			$4Na + TiCl_4 \rightarrow 4NaCl + Ti$		

Phosphorus reacts with chlorine to produce phosphorus trichloride. In one of these reactions, 6.2 g of P_4 reacts with 21.3 g of Cl_2 to produce 27.5 g of PCl_3. Determine the balanced symbol equation for this reaction. Relative formula masses (M_r): P_4 = 124, Cl_2 = 71, PCl_3 = 137.5 [3]

Amount of $P_4 = \frac{6.2}{124}$ = 0.05 mol   Amount of $Cl_2 = \frac{21.3}{71}$ = 0.3 mol   Amount of $PCl_3 = \frac{27.5}{137.5}$ = 0.2 mol[1]

Simplest whole number ratio (dividing all by 0.05) is 1 : 6 : 4[1]

$P_4 + 6Cl_2 \rightarrow 4PCl_3$ [1]

LIMITING REACTANTS

In a reaction between two reactants, one of them limits the amount of product.

Limiting and excess

Unless you can mix two reactants together in exactly the correct amounts according to the balanced equation:
- one reactant is described as being in **excess**
- the other reactant is the **limiting reactant**.

When magnesium reacts with dilute hydrochloric acid, magnesium is the limiting reactant if it all gets used up and some acid is left behind at the end.

The amount or mass of product is directly proportional to the amount or mass of the limiting reactant.

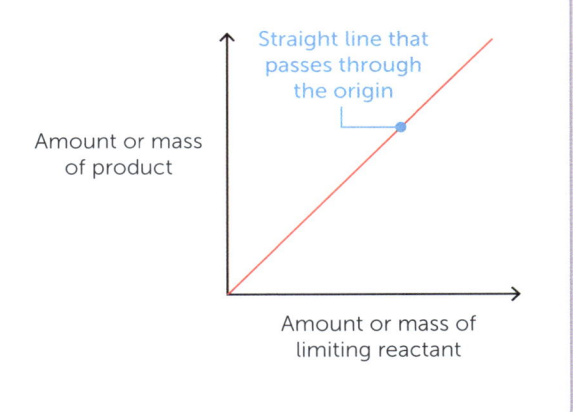

Determining the limiting reactant

You can determine which reactant is limiting in a reaction if you know:
- the balanced symbol equation
- the masses of the two reactants, and
- the relative formula mass of these two substances.

For example, 675 g of aluminium powder is added to 1.6 kg of iron(III) oxide powder and heated. The equation for the reaction is: $2Al + Fe_2O_3 \rightarrow Al_2O_3 + 2Fe$

Explain why iron(III) oxide is the limiting reactant. Relative masses: Al = 27, Fe_2O_3 = 160

1. Work out the amount in moles of each reactant:

   amount of Al = $\frac{675}{27}$ = 25 mol   amount of Fe_2O_3 = $\frac{1.6 \times 1000}{160}$ = 10 mol

2. Using your answer to step 1 and the balanced equation, check whether the amount of aluminium is enough to react with all the iron(III) oxide.

The mole ratio of Al : Fe_2O_3 is 2 : 1, so (2 × 10) = 20 mol of aluminium is needed. This is less than the amount used, so aluminium is in excess and iron(III) oxide must be the limiting reactant.

0.69 g of sodium is heated with 0.71 g of chlorine. The equation for the reaction is:

$$2Na + Cl_2 \rightarrow 2NaCl$$

Explain why sodium is in excess. Relative atomic masses (A_r): Na = 23, Cl = 35.5   [4]

M_r of Cl_2 = 2 × 35.5 = 71[1]

Amount of Na = $\frac{0.69}{23}$ = 0.03 mol[1]   Amount of Cl_2 = $\frac{0.71}{71}$ = 0.01 mol[1]

Sodium is in excess because (2 × 0.01) = 0.02 mol of Na is needed and 0.03 mol is added.[1]

CONCENTRATION OF SOLUTIONS

The **concentration** of a solution is a measure of how much **solute** a given volume contains.

Volume

The volume of a substance can be measured in m^3 but this unit is too large for laboratory chemistry. Instead, use cubic decimetres, (dm^3) or cubic centimetres (cm^3).

Measuring cylinders and other laboratory apparatus are often graduated in ml:

$$1\ ml = 1\ cm^3$$

$1\ dm^3 = 1000\ cm^3$. Divide by 1000 to convert from cm^3 (or ml) to dm^3.

For example, $80\ ml = 80\ cm^3 = 80 / 1000 = 0.08\ dm^3$.

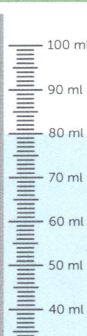

Concentration in terms of mass

You can calculate the mass of solute dissolved in a given volume of a **solution** if you know the concentration in g/dm^3 (grams per dm^3):

$$\text{mass of solute (g)} = \text{concentration of solution } (g/dm^3) \times \text{volume of solution } (dm^3)$$

Higher Tier only

The concentration of solution is greater when:
- a greater mass of solute is dissolved in a given volume of solution
- a given mass of solute is contained in a smaller volume of solution.

 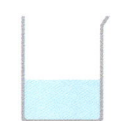 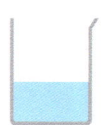 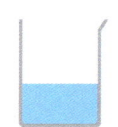 

Increasing concentration of solution

---

1. A solution of copper(II) sulfate has a concentration of 25 g/dm^3. Calculate the mass of copper(II) sulfate in 100 cm^3 of this solution. [2]

2. **Higher Tier only:** A student wants to make a 10 g/dm^3 solution of sodium chloride using 5 g of NaCl or 0.2 dm^3 of water. Determine **two** ways to make this solution. [4]

   1. Volume = $\frac{100}{1000}$ = 0.1[1] Mass = 25 × 0.1 = 2.5 g.[1] Correct answer scores both marks.[2]

   2. Dissolve 2 g of NaCl in water to make 0.2 dm^3 of solution[1], $\frac{2\ g}{0.2\ dm^3}$ = 10 g/dm^3[1]

   Dissolve 5 g of NaCl in water to make 0.5 dm^3 of solution[1], $\frac{5\ g}{0.5\ dm^3}$ = 10 g/dm^3[1]

4.3.3.1 Chemistry

PERCENTAGE YIELD

The **yield** of a reaction is the mass of a product obtained in a reaction. An efficient chemical process will have a high percentage yield.

Losing product

The maximum possible yield of a reaction, taking into account the balanced symbol equation and the masses of reactants, is the **theoretical yield**. The actual yield is often less than this because:
- a **reversible** reaction may not go to completion
- some of the product may remain in the apparatus during separation and purification
- side-reactions may occur, with some of the reactants reacting in unexpected ways.

You can calculate the **percentage yield** of a reaction using this equation:

$$\text{Percentage yield} = \frac{\text{actual yield}}{\text{theoretical yield}} \times 100$$

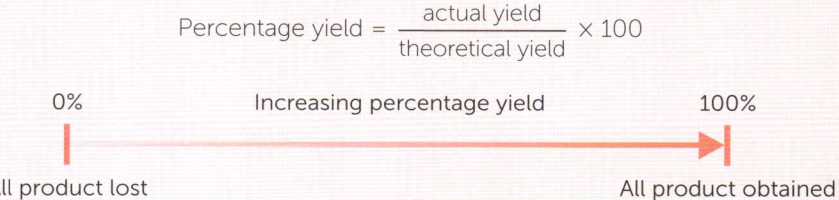

0% — All product lost — Increasing percentage yield — 100% All product obtained

Calculating theoretical yield — Higher Tier only

You can calculate the theoretical yield of a reaction if you know:
- the balanced symbol equation
- the mass of the limiting reactant, and
- the relative masses of the limiting reactant and desired product.

The calculations involved are similar to the ones explained on **page 49**.

> ★ The percentage yield cannot be greater than 100%. If your calculations show this, you have made a mistake such as inputting the masses the wrong way round.

---

1. A student extracts copper by heating copper(II) oxide with carbon. The student calculates that 1.45 g of copper should be obtained, but only obtains 0.87 g of copper. Calculate the percentage yield. [1]

2. **Higher Tier only:** Chlorine is manufactured by passing electricity through concentrated aqueous sodium chloride: $2NaCl(aq) + 2H_2O(l) \rightarrow 2NaOH(aq) + H_2(g) + Cl_2(g)$

   Calculate the theoretical yield of chlorine if 8.0 kg of sodium chloride is used.
   Relative atomic masses (A_r): H = 1, O = 16, Na = 23, Cl = 35.5 [4]

---

1. Percentage yield = $\frac{0.87\text{ g}}{1.45\text{ g}} \times 100 = 60\%$ [1]

2. M_r of NaOH = 23 + 16 + 1 = 40, M_r of Cl_2 = (2 × 35.5) = 71 [1]

   Amount of NaOH = $\frac{8.0 \times 1000\text{ g}}{40}$ = 200 mol [1]

   From the balanced equation, mole ratio NaOH : Cl_2 = 2 : 1 so $\frac{200}{2} \times 1$ = 100 mol of Cl_2 [1]

   Theoretical yield = 71 × 100 = 7100 g [1] (7.1 kg). Correct answer scores full marks. [4]

ATOM ECONOMY

Atom economy (or utilisation) is a measure of the amount of reactants that become a useful product in a chemical reaction.

Calculating atom economy

You can calculate the percentage atom economy of a reaction using the balanced symbol equation and this equation:

$$\text{Percentage atom economy} = \frac{\text{Total } M_r \text{ of the desired product}}{\text{Total } M_r \text{ of all the reactants}} \times 100$$

Sustainable development involves making sure that the things we do today will not negatively impact future generations or prevent them from enjoying the same sort of things that we have. Chemical processes with high atom economies are important for sustainable development because they use fewer resources and produce less waste. They are also more economically viable than processes with low atom economies.

Reaction pathways — Higher Tier only

You should be able to explain why a particular chemical process is used to make a given product. The factors involved include:

- the percentage yield and atom economy
- the rate of the reactions *(see pages 88–89)*
- the equilibrium position of any reversible reactions *(see pages 96–101)*
- how useful any **by-products** may be.

These factors are not necessarily linked. For example, a process may have a high percentage yield but a low percentage atom economy. Remember that some by-products may be toxic or damaging to the environment. A pathway that forms such a substance is less desirable than a pathway that does not.

Copper can be extracted by heating copper(II) carbonate with carbon:

$$2CuCO_3 + C \rightarrow 2Cu + 3CO_2$$

(a) Calculate the percentage atom economy of this process. [3]
    Relative masses: C = 12, Cu = 63.5, $CuCO_3$ = 123.5

(b) **Higher Tier only:** Suggest a way in which the atom economy of this process could be improved. Give a reason for your answer. [2]

(a) Total mass of reactants = (2 × 123.5) + 12 = 259,
total mass of desired product = (2 × 63.5) = 127.[1]
percentage atom economy = $\frac{127}{259}$ × 100 [1] = 49%.[1]

(b) Sell the carbon dioxide by-product, for example to make fizzy drinks.[1] This would make all the products desired products, so the atom economy would be 100%.[1]

4.3.4 Chemistry — Higher Tier

CONCENTRATIONS IN MOL / DM³

The concentrations of solutions can be given in mol/dm³, not just in g/dm³.

Concentration calculations

You can calculate the concentration of a solution using this equation:

$$\text{concentration (mol/dm}^3) = \frac{\text{amount of solute (mol)}}{\text{volume of solution (dm}^3)}$$

This equation can also be used to calculate the amount of solute if you know the volume and concentration of the solution. For example, 0.200 dm³ of 0.125 mol/dm³ sodium hydroxide solution contains 0.0125 mol of sodium hydroxide:

$$0.125 \text{ mol/dm}^3 = \frac{\text{amount of solute (mol)}}{0.200 \text{ dm}^3}$$

amount of sodium hydroxide = 0.125 × 0.200 = 0.025 mol

You can calculate the mass of solute in g if you know its relative formula mass. M_r of NaOH = 40, so the mass of sodium hydroxide in the solution above is (0.025 × 40) = 1.0 g.

Sodium hydroxide is a corrosive white solid that dissolves in water to form a colourless, alkaline solution.

Reacting solutions

When two solutions react together, you can calculate the concentration of one of the solutions if you know:
- the concentration of the other solution
- the reacting volumes of the two solutions.

This is useful when carrying out a **titration** between an acid and an alkali. You can revise titration calculations on **page 71**.

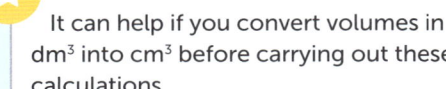

It can help if you convert volumes in dm³ into cm³ before carrying out these calculations.

1. A teacher prepares 250 cm³ of 0.4 mol/dm³ ammonium nitrate solution. Calculate the mass of ammonium nitrate the teacher used.

   Relative formula mass (M_r) of NH_4NO_3 = 80   [2]

2. Sulfuric acid reacts with sodium hydroxide:

   $$H_2SO_4 + 2NaOH \rightarrow Na_2SO_4 + 2H_2O$$

   In an experiment, 12 cm³ of 0.10 mol/dm³ sulfuric acid reacts exactly with 25 cm³ of sodium hydroxide solution.

   Calculate the concentration of the sodium hydroxide solution.   [3]

---

1. Amount of NH_4NO_3 = $\frac{250}{1000}$ × 0.4 = 0.1 mol [1]

   Mass = 0.1 × 80 = 8 g [1]

2. Volume of H_2SO_4 = 12/1000 = 0.012 dm³, volume of NaOH = 25/1000 = 0.025 dm³

   Amount of sulfuric acid = 0.10 mol/dm³ × 0.012 dm³ = 0.0012 mol. [1]

   From the balanced equation, mole ratio H_2SO_4 : NaOH is 1 : 2, so 2 × 0.0012 = 0.0024 mol of NaOH is used. [1]

   Concentration of NaOH = $\frac{0.0024 \text{ mol}}{0.0025 \text{ dm}^3}$

   = 0.096 mol/dm³ [1]

AQA GCSE **Chemistry** 8462 / 8464 – Topic 3

4.3.5 Chemistry — Higher Tier

VOLUMES OF GASES

The volume of a given amount of gas depends on the temperature and pressure.

Reacting gases

At a given temperature and pressure, equal volumes of gases contain the same amount of molecules. This means that the volume of 1 mol of hydrogen is the same as 1 mol of oxygen, provided they are under the same conditions. This means you can calculate the volume of a gas in a reaction if you know the volume of another gas in the reaction and the balanced equation.

Hydrogen reacts with chlorine to produce hydrogen chloride: $H_2(g) + Cl_2(g) \rightarrow 2HCl(g)$

If you have 50 cm³ of H_2, it will react with 50 cm³ of Cl_2 to produce (2 × 50) = 100 cm³ of HCl.

Molar gas volume

Room temperature and pressure is taken as 20 °C and 1 atmosphere. Under these conditions, 1 mol of any gas occupies 24 dm³. It does not matter what gas it is. For example, 1 mol of hydrogen occupies 24 dm³ and so does 1 mol of chlorine. Two moles of hydrogen chloride occupy (2 × 24) = 48 dm³ at room temperature and pressure.

You can calculate the volume occupied by a given amount of gas if you know:
- the mass of the gas
- the relative formula mass of the gas.

Volume at room temperature and pressure (dm³) = $\dfrac{\text{mass (g)}}{M_r \text{ of gas}} \times 24 \text{ dm}^3$

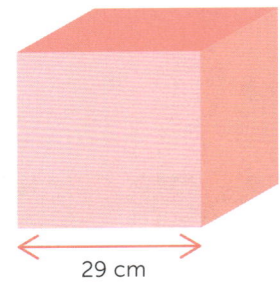

24 dm³ is approximately the volume of a cube with side length 29 cm.

---

1. 150 cm³ of hydrogen reacts with oxygen: $2H_2(g) + O_2(g) \rightarrow 2H_2O(l)$
   Determine the volume of oxygen used in this reaction. [1]
2. Calculate the volume occupied by 0.45 g of helium at room temperature and pressure. [2]
   Relative atomic mass of helium = 4
3. Calcium carbonate decomposes when heated: $CaCO_3(s) \rightarrow CaO(s) + CO_2(g)$
   Calculate the volume of gas formed when 1.0 g of calcium carbonate decomposes.
   Relative formula mass of $CaCO_3$ = 100 [2]

1. $\dfrac{150 \text{ cm}^3}{2}$ = 75 cm³ [1]
2. Amount of helium = $\dfrac{0.45}{4}$ = 0.1125 mol [1] Volume = 0.1125 × 24 = 2.7 dm³ [1]
3. Amount of calcium carbonate = $\dfrac{1.0}{100}$ = 0.01 mol [1]
   From the balanced equation, mole ratio $CaCO_3 : CO_2$ = 1 : 1 so 0.01 mol of CO_2 forms [1]
   Volume of CO_2 = 0.01 × 24 = 0.24 dm³ [1] (Or 240 cm³)

TOPIC 3

EXAMINATION PRACTICE

01 Calculate the relative formula mass (M_r) of the following compounds.
    01.1 Magnesium nitride, Mg_3N_2 [1]
    01.2 Iron(II) nitrate, $Fe(NO_3)_2$ [1]
    Relative atomic masses (A_r): N = 14, O = 16, Mg = 24, Fe = 56

02 Carbonic acid, H_2CO_3, is found naturally in rainwater.
    Calculate the percentage by mass of oxygen in carbonic acid.
    Relative atomic mass (A_r) of oxygen = 16, relative formula mass (M_r) of H_2CO_3 = 62 [2]

03 Copper(II) carbonate decomposes when it is heated:

$$CuCO_3(s) \rightarrow CuO(s) + CO_2(g)$$

    A boiling tube containing a sample of copper(II) carbonate was heated. Give a reason that explains why the mass of the boiling tube and its contents changed from 5.5 g to 4.9 g. [1]

04 A solution of sodium chloride had a concentration of 8 g/dm³.
    Calculate the mass of sodium chloride in 50 cm³ of this solution. [2]

Higher Tier only

05 Calculate the mass of 0.25 mol of carbon dioxide.
    Relative formula mass (M_r) = 44 [1]

06 Chlorine is manufactured by passing an electric current through concentrated sodium chloride solution:

$$2NaCl + 2H_2O \rightarrow 2NaOH + H_2 + Cl_2$$

    Calculate the theoretical mass of chlorine that can be made from 7.25 g of sodium chloride.
    Relative atomic masses (A_r): Na = 23, Cl = 35.5 [4]

07 Magnesium powder reacts with iron(III) oxide powder:

$$3Mg + Fe_2O_3 \rightarrow 3MgO + 2Fe$$

    960 g of magnesium was added to 2.0 kg of iron(III) oxide.
    Explain why iron(III) oxide was the limiting reactant.
    Relative atomic masses (A_r): O = 16, Mg = 24, Fe = 56 [4]

08 Red lead oxide, Pb_3O_4, is used in some rustproof paints. It can be prepared by heating lead(II) oxide, PbO, with oxygen. In one of these reactions, 6.69 g of PbO reacts with an excess of oxygen to produce 6.85 g of Pb_3O_4.
    08.1 Calculate the mass of oxygen used in this reaction. [1]
    08.2 Determine the balanced symbol equation for this reaction.
    Relative formula masses (M_r): O_2 = 32, PbO = 223, Pb_3O_4 = 685 [3]

Chemistry only

09 An iron and steel manufacturer calculated that 1 kg of iron(III) oxide should produce 700 g of iron. The manufacturer actually obtained 0.42 kg of iron from 1 kg of iron(III) oxide.
Calculate the percentage yield of iron. [2]

10 Hydrogen can be obtained in two ways using electrolysis:
Process 1: $2H_2O \rightarrow 2H_2 + O_2$
Process 2: $2NaCl + 2H_2O \rightarrow 2NaOH + Cl_2 + H_2$

  10.1 Calculate the percentage atom economy for Process 1.
  Relative formula masses (M_r): H_2O = 18, H_2 = 2, O_2 = 32 [3]

  10.2 The atom economy for Process 2 is 1.3%.
  Give a reason why the atom economies of Process 1 and Process 2 are different. [1]

  10.3 Suggest **one** way to improve the atom economy of Process 2. [1]

Higher Tier only

11 A student used 2.5 g of sodium hydroxide to make 250 cm³ of sodium hydroxide solution.
Calculate the concentration of sodium hydroxide in mol/dm³.
Relative formula mass (M_r) of NaOH = 40 [3]

12 Lead(II) nitrate solution reacted with potassium iodide solution:

$$Pb(NO_3)_2 + 2KI \rightarrow PbI_2 + 2KNO_3$$

In an experiment, 10 cm³ of a potassium iodide solution reacts exactly with 25 cm³ of 0.01 mol/dm³ lead(II) nitrate solution.
Calculate the concentration of the potassium iodide solution. [3]

13 Sulfur dioxide reacts with oxygen to produce sulfur trioxide:

$$2SO_2(g) + O_2(g) \rightarrow 2SO_3(g)$$

Calculate the volume of sulfur trioxide that can be produced from 50 cm³ of oxygen and an excess of sulfur dioxide. [1]

14 During a reaction, a student collected 84 cm³ of carbon dioxide.
  14.1 Calculate the amount in moles of carbon dioxide that the student collected. [3]
  14.2 In a separate experiment, the student collected 0.11 g of carbon dioxide.
  Calculate the volume of carbon dioxide collected.
  Relative formula mass (M_r) of carbon dioxide = 44
  The volume of one mole of any gas at room temperature and pressure is 24 dm³. [1]

OXIDATION AND REDUCTION

Gain or loss of oxygen

Oxidation and reduction reactions often involve oxygen:
- **Oxidation** is gain of oxygen
- **Reduction** is loss of oxygen.

You can work out whether a substance is oxidised or reduced in a reaction by looking at the balanced equation for the reaction. For example, copper(II) oxide reacts with carbon:
- Carbon gains oxygen and is oxidised
- Copper oxide loses oxygen and is reduced

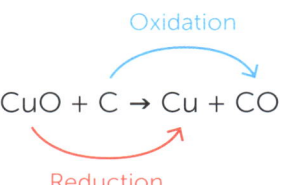

$CuO + C \rightarrow Cu + CO$

Oxidation / Reduction

Loss or gain of electrons — Higher Tier only

Oxidation and reduction is also described in terms of electrons:
- Oxidation is loss of electrons
- Reduction is gain of electrons

These definitions are useful for identifying oxidation and reduction in reactions that do not involve oxygen.

For example, sodium reacts with chlorine: $2Na + Cl_2 \rightarrow 2NaCl$

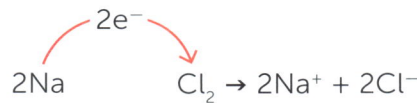

$2Na \quad Cl_2 \rightarrow 2Na^+ + 2Cl^-$

Two half equations describe what happens in this reaction:
- Sodium loses electrons and is oxidised to sodium ions: $\quad 2Na \rightarrow 2Na^+ + 2e^-$
- Chlorine gains electrons and is reduced to chloride ions: $\quad Cl_2 + 2e^- \rightarrow 2Cl^-$

Rusting is an oxidation reaction

---

1. Explain why the reactions of metals with oxygen are described as oxidation reactions. [2]
2. When powdered and heated, magnesium reacts with zinc oxide: $Mg + ZnO \rightarrow MgO + Zn$

    Identify oxidation and reduction in this reaction. Give reasons for your answer. [2]
3. **Higher Tier only:** Bromine displaces iodine from sodium iodide. The reaction can be described by two half equations:

    $Br_2 + 2e^- \rightarrow 2Br^-$ and $2I^- \rightarrow I_2 + 2e^-$

    Explain which substance is oxidised and which substance is reduced. [2]

    1. Oxidation is gain of oxygen[1] and metals gain oxygen when they become metal oxides[1].
    2. Magnesium gains oxygen and is oxidised to magnesium oxide[1], zinc oxide loses oxygen and is reduced to zinc[1].
    3. Bromine is reduced because it gains electrons to form bromide ions.[1] Iodide ions are oxidised because they lose electrons to form iodine.[1]

THE REACTIVITY SERIES

The **reactivity series** lists metals in order of their reactivity.

Losing electrons

Metals lose **electrons** to form positively charged **ions** when they react with other substances. In general, the more easily a metal forms positively charged ions, the more **reactive** it is.

The reactivity series is based on observing reactions with water, dilute acids and other substances. The more reactive the metal, the more vigorous the reaction.

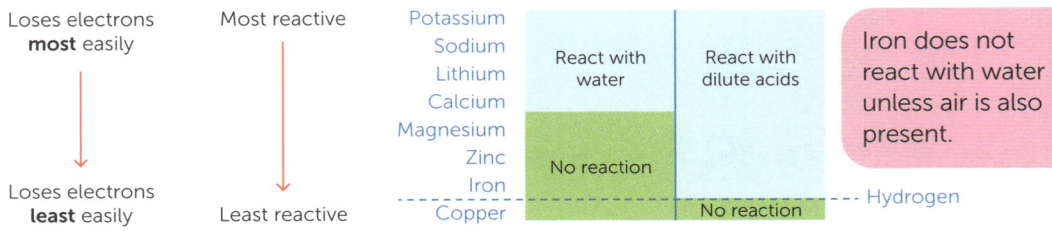

Reactions with water and dilute acids

Hydrogen is given off if a metal reacts with water or with a dilute acid. The more reactive the metal, the faster the gas is given off. You can revise the reactions of potassium, sodium and lithium with water on **page 19**, and the reactions of metals with dilute acids on **page 62**. Hydrogen is a non-metal element but is often included in the reactivity series. This is because metals that are more reactive than hydrogen can react with dilute acids.

Displacement reactions

A more reactive metal can **displace** a less reactive metal from its compounds. For example, a red-brown coating of copper forms when a piece of zinc is dipped into copper(II) sulfate solution. Zinc ions replace copper ions in the solution because zinc is more reactive than copper:

$$Zn(s) + CuSO_4(aq) \rightarrow ZnSO_4(aq) + Cu(s)$$

1. Describe what happens when an iron nail is placed in water. [2]
2. Compare the reactions of zinc and iron with dilute hydrochloric acid. [4]
3. Explain why magnesium reacts with iron(III) oxide, but iron does not react with magnesium oxide. [2]

   1. It rusts if air or oxygen is present[1] forming orange-brown hydrated iron(III) oxide[1].
   2. Both reactions produce hydrogen[1] but zinc gives steady bubbling and iron gives slow bubbling[1]. Both reactions produce metal chlorides[1] but the reaction with zinc produces zinc chloride and the reaction with iron produces iron(II) chloride[1].
   3. Magnesium is more reactive than iron.[1] This means that magnesium can displace iron from iron(III) oxide, but iron is not reactive enough to displace magnesium from magnesium oxide.[1]

EXTRACTION OF METALS

Gold and other unreactive metals may be found in the Earth's crust as **native metals**. This is the pure form of the metals rather than one of their compounds. However, most metals are found combined with other elements such as oxygen. These metals must be extracted using chemical reactions. **Ores** contain enough of a metal or its compound for extraction to be profitable.

Extraction and the reactivity series

The method used to extract a metal is related to its position on the reactivity series:

- Metals that are more reactive than carbon must be extracted by electrolysis of one of their compounds.
- Metals that are less reactive than carbon can be extracted by heating with carbon.

The four metals shown in **black** here are commonly used metals, but you do not need to recall what happens when they are added to water or dilute acids.

Chemical reactions may be needed to purify silver, gold and platinum, but not to extract them from an ore.

Remember that carbon is a non-metal, not a metal.

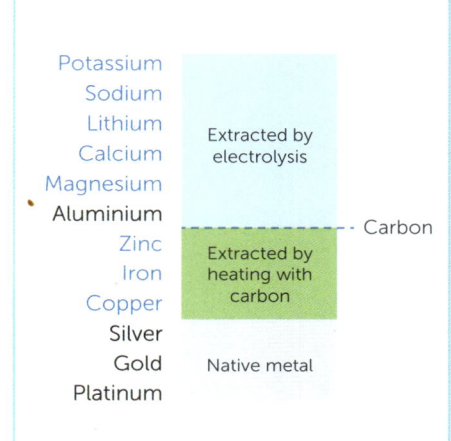

Reduction of oxides

Ores often contain metal oxides, or compounds that can be converted into metal oxides. If a metal is less reactive than carbon, its oxide can be reduced by heating with carbon. Zinc is a metal that can be extracted this way:

- Zinc oxide is reduced to zinc
- Carbon is oxidised to carbon monoxide

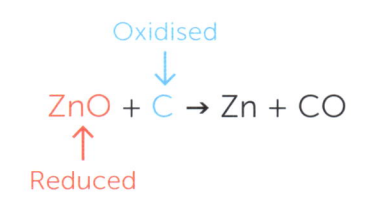

---

1. Give a reason why aluminium cannot be extracted from its oxide by heating with carbon. [1]
2. Haematite is an iron ore that contains iron(III) oxide, Fe_2O_3.
   Suggest a method to extract iron from haematite. Explain your answer. [3]
3. **Higher Tier only:** Rubidium is produced by reacting sodium with molten rubidium chloride:

   $$Na + RbCl \rightarrow NaCl + Rb$$

   Explain why rubidium is reduced in this reaction. Include a half equation in your answer. [3]

   1. Aluminium is more reactive than carbon.[1]
   2. Heat the haematite with carbon.[1] This will reduce iron(III) oxide to iron[1] because iron is less reactive than carbon[1].
   3. Reduction is gain of electrons[1] and rubidium ions gain electrons to form rubidium atoms in this reaction[1]. $Rb^+ + e^- \rightarrow Rb$.[1]

REACTIONS OF ACIDS WITH METALS

Hydrochloric acid and sulfuric acid

Metals react with **acids** to produce salts and hydrogen. The reactions with magnesium are the most vigorous and the reactions with iron are the least vigorous. The more reactive the metal, the greater the rate of bubbling.

Metal	Salts produced with:	
	Hydrochloric acid	Sulfuric acid
Magnesium	magnesium chloride (colourless solution)	magnesium sulfate (colourless solution)
Zinc	zinc chloride (colourless solution)	zinc sulfate (colourless solution)
Iron	iron(II) chloride (green solution)	iron(II) sulfate (green solution)

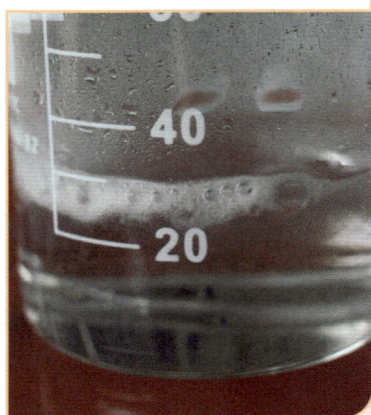

Solutions containing Fe^{2+} ions slowly turn orange-brown as these ions oxidise to Fe^{3+} ions in air.

Redox reactions — Higher Tier only

Acids in aqueous solution are a source of hydrogen ions:

$$HCl(aq) \rightarrow H^+(aq) + Cl^-(aq) \qquad H_2SO_4(aq) \rightarrow 2H^+(aq) + SO_4^{2-}(aq)$$

The reactions of metals with dilute acids involve metal atoms, and hydrogen ions from the acid. They are examples of redox reactions – reduction and oxidation happen at the same time:
- Hydrogen ions are reduced to hydrogen: $2H^+(aq) + 2e^- \rightarrow H_2(g)$
- Metal atoms are oxidised to metal ions: $M(s) \rightarrow M^{2+}(aq) + 2e^-$ (M stands for the metal)

You can combine these two half equations to obtain an ionic equation for the reaction:

$$M(s) + 2H^+(aq) \rightarrow M^{2+}(aq) + H_2(g)$$

Overall, electrons are transferred from metal atoms to hydrogen ions in the reaction.

---

1. (a) Describe what you would see when magnesium ribbon reacts with excess dilute hydrochloric acid. [3]
   (b) Complete the balanced equation for this reaction. Include state symbols. [2]
   $$Mg(\_\_\_) + \_\_\_ HCl(\_\_\_) \rightarrow MgCl_2(\_\_\_) + H_2(\_\_\_)$$
2. **Higher Tier only:** Explain why the reaction of zinc with dilute sulfuric acid is a redox reaction. [4]

   1. (a) Rapid bubbling.[1] The magnesium gradually disappears[1] to form a colourless solution[1].
      (b) $Mg(s) + 2HCl(aq) \rightarrow MgCl_2(aq) + H_2(g)$ Correctly balanced[1], correct state symbols[1].
   2. Hydrogen ions from the acid gain electrons[1] and are reduced to hydrogen[1]. Zinc atoms lose electrons[1] and are oxidised to zinc ions[1].

4.4.2.2 | 5.4.2.2

FORMULAE OF IONIC COMPOUNDS

You can work out the **formula** of an ionic compound if you know the formulae of the **ions** it contains.

Formulae of common ions

The table shows the formulae of some common ions.

Positive ions	
Name	Formula
Ammonium	NH_4^+
Hydrogen	H^+
Lithium	Li^+
Potassium	K^+
Silver	Ag^+
Sodium	Na^+
Barium	Ba^{2+}
Calcium	Ca^{2+}
Copper(II)	Cu^{2+}
Magnesium	Mg^{2+}
Zinc	Zn^{2+}
Iron(II)	Fe^{2+}
Lead(II)	Pb^{2+}
Iron(III)	Fe^{3+}
Aluminium	Al^{3+}

Negative ions	
Name	Formula
Bromide	Br^-
Chloride	Cl^-
Hydrogencarbonate	HCO_3^-
Hydroxide	OH^-
Iodide	I^-
Nitrate	NO_3^-
Carbonate	CO_3^{2-}
Oxide	O^{2-}
Sulfate	SO_4^{2-}
Sulfide	S^{2-}
Phosphate	PO_4^{3-}

HCO_3^- is the hydrogencarbonate ion, but the names of compounds containing this ion end in 'hydrogen carbonate'.

You will be given the formulae of the ions needed to deduce the formula of an ionic compound.

Deducing formulae

There are equal numbers of positive and negative charges in an ionic compound. You need to take this into account when working out a formula. The charges are written in superscript.

- Sodium chloride contains Na^+ ions and Cl^- ions – its formula is NaCl.
- Ammonium sulfate contains NH_4^+ ions and SO_4^{2-} ions – its formula is $(NH_4)_2SO_4$.

A compound ion contains more than one element. Its formula goes inside brackets if you need two or more of the ionic compound in the formula. So $Mg(OH)_2$ is correct but $MgOH_2$ is not.

1. Write the formulae of the following ionic compounds: [7]
   (a) zinc bromide
   (b) iron(III) oxide
   (c) aluminium hydroxide
   (d) copper(II) carbonate
   (e) ammonium chloride
   (f) potassium phosphate
   (g) calcium nitrate

2. Name the following ionic compounds:
   (a) $PbCl_2$ (b) Ag_2O (c) CaS
   (d) $NaHCO_3$ (e) $Ba_3(PO_4)_2$ [5]

1. (a) $ZnBr_2$[1]   (b) Fe_2O_3[1]   (c) $Al(OH)_3$[1]
   (d) $CuCO_3$[1]   (e) NH_4Cl[1]   (f) K_3PO_4[1]
   (g) $Ca(NO_3)_2$[1]

2. (a) lead(II) chloride[1]   (b) silver oxide[1]
   (c) calcium sulfide[1]
   (d) sodium hydrogen carbonate[1]
   (e) barium phosphate[1]

NEUTRALISATION OF ACIDS

Bases, alkalis and metal carbonates

Acids are **neutralised** by bases, alkalis and carbonates. A **base** reacts with an acid to produce a salt and water only. Bases include metal hydroxides and metal oxides. **Alkalis** are soluble bases, usually metal hydroxides. **Carbonates** react with acids to produce a salt, water and carbon dioxide.

Predicting products

Neutralisation reactions always produce a salt and water. Reactions with a carbonate produce carbon dioxide as well. Work out which salt is formed using the rules below. You can write the formula of the salt if you know the formulae of the ions it contains.

Salts

A **salt** is a substance formed when hydrogen ions in an acid are replaced by other ions. The particular salt formed in a neutralisation reaction depends upon the two reactants:
- the positive metal ion or ammonium ion in the base, alkali or carbonate
- the acid used:
    - hydrochloric acid contains Cl^- ions and forms chlorides
    - nitric acid contains NO_3^- and forms nitrates
    - sulfuric acid contains SO_4^{2-} ions forms sulfates

Ammonium nitrate is used in fertilisers to improve the yields of crops.

The table gives some examples.

	Hydrochloric acid	**Nitric acid**	**Sulfuric acid**
Copper(II) oxide	copper(II) chloride $CuCl_2$	copper(II) nitrate $Cu(NO_3)_2$	copper(II) sulfate $CuSO_4$
Sodium hydroxide	sodium chloride $NaCl$	sodium nitrate $NaNO_3$	sodium sulfate Na_2SO_4
Ammonium carbonate	ammonium chloride NH_4Cl	ammonium nitrate NH_4NO_3	ammonium sulfate $(NH_4)_2SO_4$

1. Write word equations for the reactions between the following pairs of reactants.
    (a) Zinc oxide and nitric acid   [2]   (b) Potassium hydroxide and sulfuric acid   [2]
2. Write the formula for the salt formed when iron(II) carbonate reacts with nitric acid.   [1]
    Formulae of ions: Fe^{2+} and NO_3^-
3. Complete the equation for the reaction between calcium carbonate and hydrochloric acid.   [2]

$$\_\_\_\_\_ CaCO_3 + \_\_\_\_\_ HCl \rightarrow \_\_\_\_\_ + \_\_\_\_\_ + \_\_\_\_\_$$

1. (a) zinc oxide + nitric acid → zinc nitrate[1] + water[1]
   (b) potassium hydroxide + sulfuric acid → potassium sulfate[1] + water[1]
2. $Fe(NO_3)_2$ [1]
3. $CaCO_3 + 2HCl \rightarrow CaCl_2 + H_2O + CO_2$ (correct formula of reactants[1] correct formulae of products[1] correctly balanced[1]).

SALTS FROM INSOLUBLE REACTANTS

Soluble salts can be made using an acid and a suitable solid, **insoluble** reactant. The insoluble reactant can be a reactive metal, a metal oxide or metal hydroxide, or a metal carbonate. Once the reaction has finished, the excess solid is filtered off leaving a salt solution. This solution can be crystallised to produce a solid salt.

Choosing the solid reactant

The choice of reactant depends on the reactivity of the metal. You need to consider whether the metal reacts with dilute acids and, if it does, whether it is too reactive to use safely.

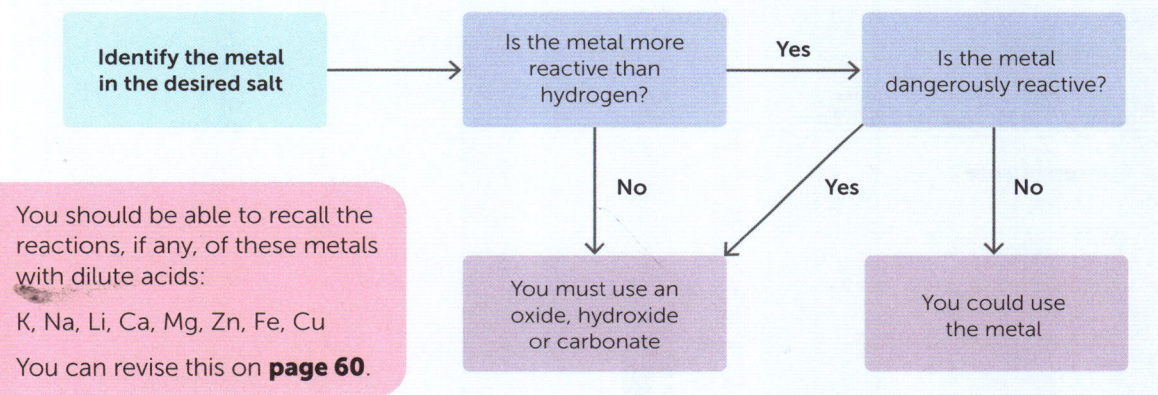

You should be able to recall the reactions, if any, of these metals with dilute acids:

K, Na, Li, Ca, Mg, Zn, Fe, Cu

You can revise this on **page 60**.

Outline of the method

Required practical 1 involves preparing a pure, dry sample of a soluble salt from an insoluble metal oxide or metal carbonate. You can revise this practical in detail on **page 66**.

In outline:

1. Add the insoluble solid to the dilute acid until no more reacts and some is left over.
2. Use filtration to remove the excess insoluble solid. Collect the filtrate (the salt solution).
3. Use crystallisation to produce the solid salt from the salt solution.

You can revise these separation methods on **pages 5 and 6**.

1. Give a reason why copper cannot be used to make copper(II) sulfate. [1]
2. Give **two** reasons why lithium, sodium and potassium salts cannot be made from insoluble reactants. [2]
3. Name **two** insoluble substances that can be used to make zinc chloride. [2]

1. Copper does not react with dilute acids.[1]
2. Lithium, sodium and potassium are too reactive to use with acids safely.[1] Their oxides, hydroxides and carbonates are all soluble in water.[1]
3. Two from: zinc[1], zinc oxide[1], zinc hydroxide[1], zinc carbonate[1].

REQUIRED PRACTICAL 1 (8)

Preparing a pure, dry soluble salt

This activity helps you develop your ability to heat substances safely, to handle them safely and carefully, and to separate and purify them.

Heating water and solutions

The **Bunsen burner** is commonly used to heat substances. Adjust the rate of heating by carefully turning the gas tap once the Bunsen burner is lit and turning the collar to adjust the size of the air hole.

The orange safety flame, obtained by closing the air hole, should not be used for heating.

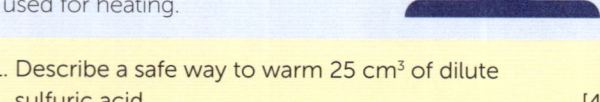

Making a salt solution

Copper(II) oxide is an insoluble black powder. It reacts with dilute sulfuric acid to produce copper(II) sulfate and water. You should add a small mass of copper(II) oxide to warm acid using a **spatula**, then stir with a glass rod. Repeat until excess copper(II) oxide has been added.

1. Describe a safe way to warm 25 cm³ of dilute sulfuric acid. [4]

    Place the acid in a 100 cm³ beaker[1]. Set up a tripod and gauze on a heat-resistant mat, and place the beaker on the gauze[1]. Heat the acid gently using a Bunsen burner[1], then turn off the Bunsen burner before the acid begins to boil[1].

2. (a) Describe how you will know that excess copper(II) oxide has been added. [2]
   (b) Explain why it is necessary to add excess copper(II) oxide. [2]

    (a) The copper(II) oxide will no longer disappear[1] but will form a cloudy mixture instead[1].

    (b) This ensures that all the sulfuric acid will have reacted[1] which makes it safer to carry out the next stages in the experiment[1].

Preparing crystals from a salt solution

Use **filtration** (page 5) to remove excess copper(II) oxide from the copper(II) sulfate solution. Then use **crystallisation** to evaporate water from the filtrate (page 6). Large, regularly shaped crystals are obtained if the water is evaporated slowly, and the solution is not heated to dryness over the boiling water bath.

3. Suggest **three** precautions needed for safe working in this activity. Give reasons for your answers. [3]

    Tie hair and loose clothing to avoid it catching fire in the Bunsen burner flame.[1] Let the beaker and its contents cool before filtering to avoid getting burned.[1] Wear eye protection and avoid skin contact because sulfuric acid and copper(II) sulfate are irritants.[1]

THE pH SCALE

The **pH scale** goes from 0 to 14. It is a measure of the acidity or alkalinity of a solution.

Acids

Acids produce hydrogen ions, H^+(aq), when they dissolve in water. The solutions formed are described as being **acidic**, and have pH values **less than 7**.

Alkalis

Alkalis dissolve in water to form solutions described as being **alkaline**. These solutions contain hydroxide ions, OH^-(aq). They have pH values **greater than 7**.

Solutions with a pH value of 7 have **equal** concentrations of H^+ ions and OH^- ions. They are described as being **neutral**.

Using indicators

An **indicator** is a substance that changes colour depending on the pH of the solution it is in. Universal indicator is a mixture of indicators intended to give a spectrum of colours. It is a wide range indicator because it can be used to estimate a range of pH values, not just one.

Neutralisation

Alkalis are soluble bases, usually metal hydroxides. When an alkali neutralises an acid:

- H^+ ions in the acidic solution react with OH^- in the alkaline solution
- water is produced in the reaction.

You can represent this neutralisation reaction with an ionic equation:

$$H^+(aq) + OH^-(aq) \rightarrow H_2O(l)$$

1. Describe how to use universal indicator paper to measure the approximate pH of a solution. [3]
2. The acidity or alkalinity of a solution can be measured using universal indicator solution or a pH probe. Compare these two methods. [4]

1. Use a clean glass rod to add a spot of the solution to the paper[1], leave for 30 seconds for the colour to develop[1] then match it to a colour on a pH colour chart[1].

2. Both methods produce a pH value[1] but the value obtained by a pH probe is more accurate[1]. Universal indicator can only produce pH values with whole numbers but a pH probe can give pH values precise to 1 or 2 decimal places.[1] A pH probe must be calibrated to give accurate readings but universal indicator solution does not.[1]

4.4.2.4 | 5.4.2.4

INVESTIGATING NEUTRALISATION

The **pH** of a reaction mixture changes as an acid is added to an alkali. Solutions of strong acids such as hydrochloric acid tend to have very low pH values. Solutions of strong alkalis such as sodium hydroxide tend to have very high pH values. A neutral solution forms if they are mixed together in just the right proportions.

An example method

You can investigate how pH changes as dilute hydrochloric acid is added gradually to sodium hydroxide solution.

1. Add 20 cm³ of sodium hydroxide solution to a beaker.
2. Stir with a glass rod, then measure the pH.
3. Add 1 cm³ of dilute hydrochloric acid using a dropping pipette.
4. Repeat steps 2 and 3 until you have added a total of 40 cm³ of hydrochloric acid.

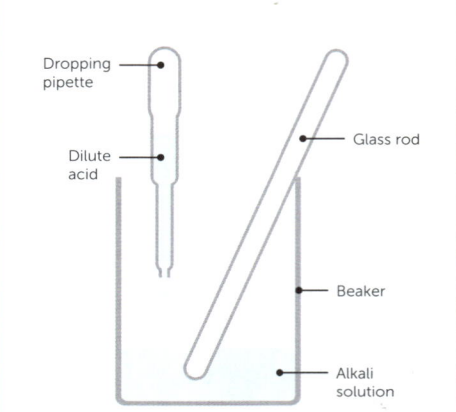

Example results

The graph shows results from an experiment like this one.

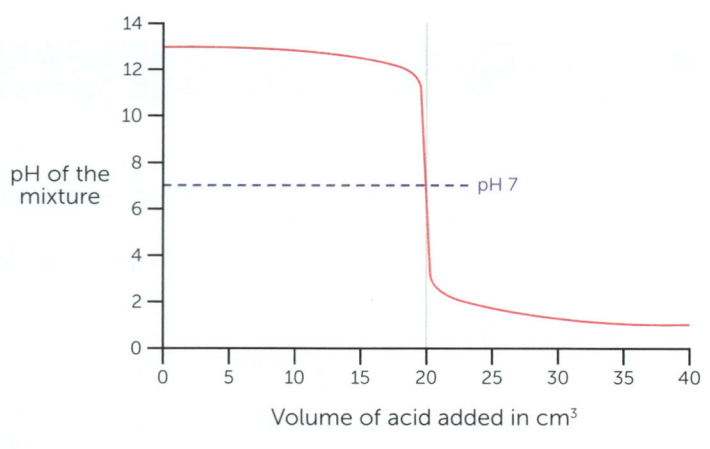

This method is the basis of carrying out a titration.
You can revise titrations on **page 71**.

1. Determine the volume of acid needed to exactly neutralise the alkali. [1]
2. Describe what is happening: (a) In the range 0–20 cm³ [2]   (b) In the range 20–40 cm³ [2]

   1. 20 cm³[1]

   2. (a) The pH falls gradually up to about 18 cm³[1] then it falls rapidly to pH 7[1].

     (b) The pH falls rapidly up to about 22 cm³[1] then it falls gradually to pH 1.5[1].

TITRATIONS

4.4.2.5 | **Chemistry** | **RPA2**

A **titration** lets you determine volumes of acid and alkali that react together. You will need a **burette** to release accurate, variable volumes of liquid and a **volumetric pipette** to deliver a particular volume in an accurate and repeatable way.

Titres

The **end point** occurs when the indicator first changes colour permanently. This happens when a strong acid (such as hydrochloric, nitric or sulfuric acid) exactly neutralises a strong alkali (such as sodium hydroxide). A **titre** is the volume of acid needed to reach the end point:

titre = (end reading on burette) − (start reading on burette)

Carrying out a titration

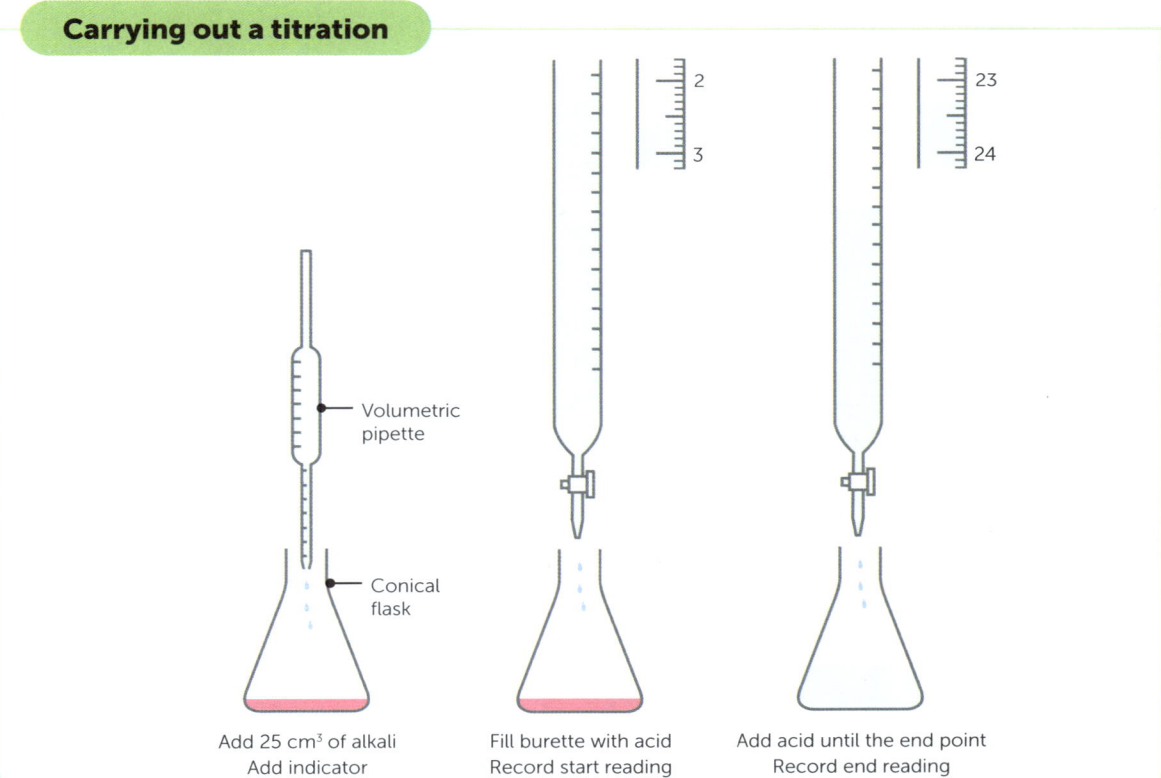

Add 25 cm³ of alkali
Add indicator

Fill burette with acid
Record start reading

Add acid until the end point
Record end reading

1. Determine the titre shown in the diagram. [2]
2. Explain why a single indicator such as phenolphthalein is used instead of universal indicator. [4]
   (a) (23.30 − 2.60)[1] = 20.70 cm³[1]
   (b) Universal indicator changes to a range of colours[1] over a range of pH values[1] but single indicators change colour over a small range of pH values[1] which gives a sharp end point[1].

AQA GCSE Chemistry 8462 / 8464 – Topic 4

Chemistry

REQUIRED PRACTICAL 2
Titration

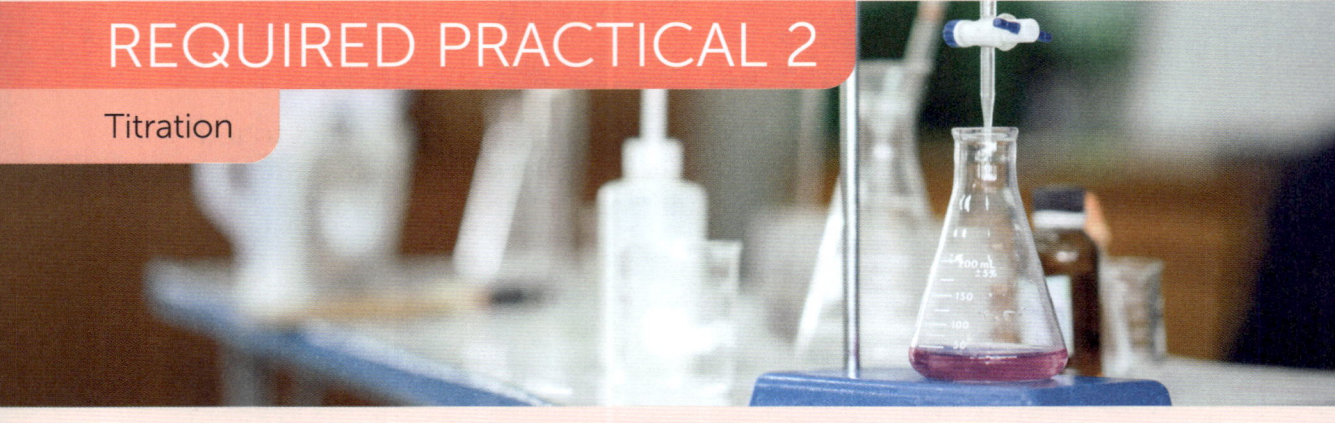

This activity helps you develop your ability to accurately make and record measurements, and to analyse unknown substances.

Outline of the practical

The aims of this practical are slightly different, depending on your tier of entry in the exam:
- Foundation – to find the volume of acid needed to neutralise a known volume of alkali
- Higher – to find the concentration of an acid solution using a known concentration of alkali.

You can revise the outline method on **page 69** and the calculations needed on **page 71**.

Precautions for accurate results and safe working

Phenolphthalein is often used as the indicator in acid-alkali titrations. It is pink in alkalis and colourless in acids.

The colour change is easier to see if the conical flask is placed on a white tile.

1. Describe **two** precautions needed to read a burette accurately. [2]
2. Describe how to record a burette reading. [3]
3. Determine the burette reading shown in the photo. [1]

   1. Read at eye level to avoid parallax errors.[1] Read to the bottom of the meniscus.[1] Place a piece of paper behind the burette to make it easier to see the meniscus.[1]
   2. Give the reading to 2 decimal places[1], ending in 0 if the meniscus is on a line and 5 if it is part-way between two lines[1].
   3. 38.10 cm^3[1]

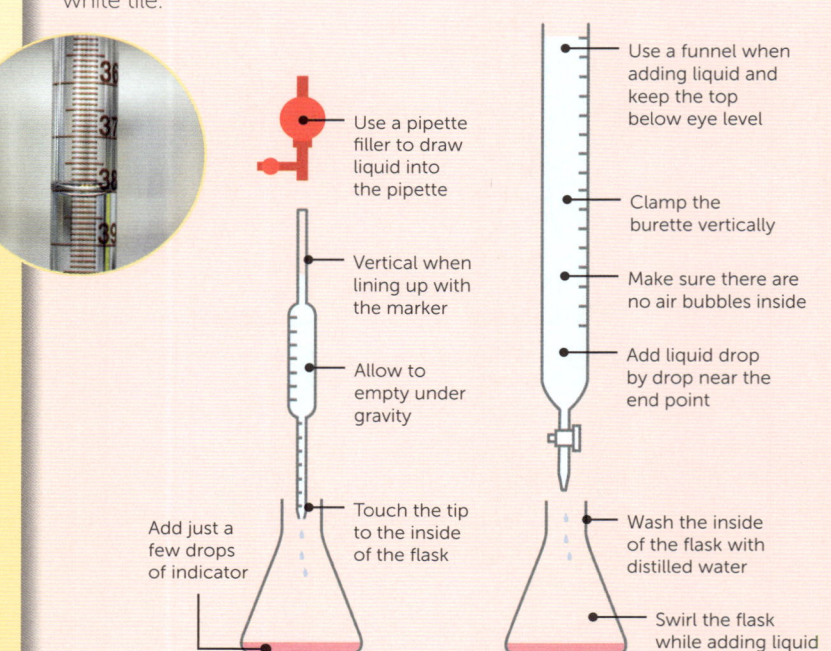

70 ClearRevise

4.4.2.5 Chemistry — Higher Tier

TITRATION CALCULATIONS

You can use titration results to calculate the concentrations of acids and alkalis.

Calculating concentrations in mol/dm³

You can use titres to calculate the concentration of an acid or alkali if you know:
- the balanced symbol equation for the reaction between the acid and alkali
- the volume and concentration of one of the solutions
- the volume of the other solution.

If there are titres from several runs, you need to calculate the mean of the **concordant** titres. These are titres that are within 0.10 cm³ of each other, such as titrations 1, 2 and 4 below:

It makes it easier to calculate the mean titre if you tick the concordant titres in your results table.

	Titration 1	Titration 2	Titration 3	Titration 4
Titre in cm³	22.55 ✓	22.50 ✓	22.65	22.45 ✓

$$\text{mean titre (cm}^3\text{)} = \frac{(22.55 + 22.50 + 22.45)}{3} = 22.50 \text{ cm}^3$$

If you know the relative formula mass of a reactant in a titration, you can carry out titration calculations for its concentration given in g/dm³. You can also convert a concentration in mol/dm³ into a concentration in g/dm³.

---

1. A student titrated 25.0 cm³ portions of 0.180 mol/dm³ sodium hydroxide solution with a dilute solution of hydrochloric acid: NaOH + HCl → NaOH + H₂O.
The mean titre was 22.50 cm³.
Calculate the concentration of the hydrochloric acid in mol/dm³. [3]

2. (a) Calculate the concentration in g/dm³ of 0.200 mol/dm³ HCl. [2]
Relative formula mass (M_r) of HCl = 36.5
(b) Calculate the mass of sodium hydroxide in 50.0 cm³ of a 0.180 mol/dm³ solution. [2]
Relative formula mass (M_r) of NaOH = 40

1. Calculate the amount of sodium hydroxide (because you know its concentration and volume):
moles NaOH = $\frac{25.0}{1000}$ × 0.180 = 0.0045 mol.[1]
Looking at the balanced equation, mol ratio NaOH : HCl = 1 : 1, so 0.0045 mol of HCl.[1]
Calculate the concentration of hydrochloric acid (because you now know its amount and volume):
concentration (mol/dm³) = $\frac{0.0045}{22.50}$ × 1000 = 0.200 mol/dm³ [1]
2. (a) Concentration (g/dm³) = 36.5 × 0.200 mol/dm³ [1] = 7.30 g/dm³ [1]
(b) Amount of NaOH = 0.180 mol/dm³ × $\frac{50.0 \text{ cm}^3}{1000}$ = 0.009 mol. [1]
Mass of NaOH = 40 × 0.009 = 0.36 g.[1]

AQA GCSE **Chemistry 8462 / 8464** – Topic 4

4.4.2.6 | Higher Tier | 5.4.2.5

STRONG AND WEAK ACIDS

Acids can be described as strong or weak, depending on how much they ionise in aqueous solution.

Strong acids

Strong acids fully ionise in aqueous solution – each acid molecule breaks down in the water to form ions. Common strong acids include:
- Hydrochloric acid: $HCl(aq) \rightarrow H^+(aq) + Cl^-(aq)$
- Sulfuric acid: $H_2SO_4(aq) \rightarrow 2H^+(aq) + SO_4^{2-}(aq)$

Weak acids

Weak acids only partially ionise in aqueous solution – only a few of their acid molecules break down in the water to form ions. Common weak acids include:
- Ethanoic acid: $CH_3COOH(aq) \rightleftharpoons H^+(aq) + CH_3COO^-(aq)$
- Carbonic acid: $H_2CO_3(aq) \rightleftharpoons 2H^+(aq) + CO_3^{2-}(aq)$

The equations for the strong acids have the usual arrow. The equations for the weak acids have a split arrow because these reactions are **reversible** and do not go to completion (see **page 96**).

Atmospheric carbon dioxide dissolves in rainwater to form carbonic acid. This seeps through the ground and dissolves limestone rocks, forming caves.

Concentration of H+ and pH

The concentration of hydrogen ions in a solution determines its pH. The greater the hydrogen ion concentration, the lower the pH.

In neutral solutions:
concentration of $H^+(aq)$ = concentration of $OH^-(aq)$

As the hydrogen ion concentration increases by a factor of 10, the pH decreases by 1. The table gives some examples of how this works for nitric acid (a strong acid) and citric acid (a weak acid).

Concentration of acid in mol/dm³	pH of nitric acid	pH of citric acid
0.01	2	2.62
0.1	1	2.08
1	0	1.57

In the exam, you only need to be able to use pH values given in whole numbers.

---

1. Explain why citric acid can be described as weak, dilute or concentrated. [3]
2. A student discovers that 5.6 mol/dm³ ethanoic acid and 0.01 mol/dm³ hydrochloric acid have the same pH value. Explain this observation. [2]

   1. Citric acid is a weak acid because it is partially dissociated in aqueous solution.[1] It is dilute when a given volume contains a small amount of acid[1] and concentrated when the volume contains a larger amount of acid[1].
   2. Ethanoic acid is a weak acid and only partially ionises in solution[1], but hydrochloric acid is a strong acid and fully ionises in solution[1]. A high enough concentration of ethanoic acid produces the same H+ concentration as a lower concentration of hydrochloric acid.[1]

72 ClearRevise

4.4.3.1, 4.4.3.5 5.4.3.1, 5.4.3.5

ELECTROLYSIS

Electrolysis is a process that can decompose compounds to simpler substances.

Electrolytes and electrodes

An **electrolyte** is a liquid or solution that can conduct electricity. **Ionic compounds** are electrolytes when they are molten (in the liquid state) or when they are dissolved in water. This is because their ions are free to move and carry electric charge from place to place.

Electrodes are substances that conduct electricity into an electrolyte. They are usually graphite, or unreactive metals such as copper and platinum. **Inert** electrodes only provide a charged surface for reactions to happen on, and do not become chemically changed themselves.

Metals and ionic compounds conduct electricity when they are liquid, but only metals (and graphite) conduct when they are solid.

The process of electrolysis

Ions move to oppositely charged electrodes when an electric current passes through an electrolyte. When ions reach the electrodes, they are **discharged** as elements.

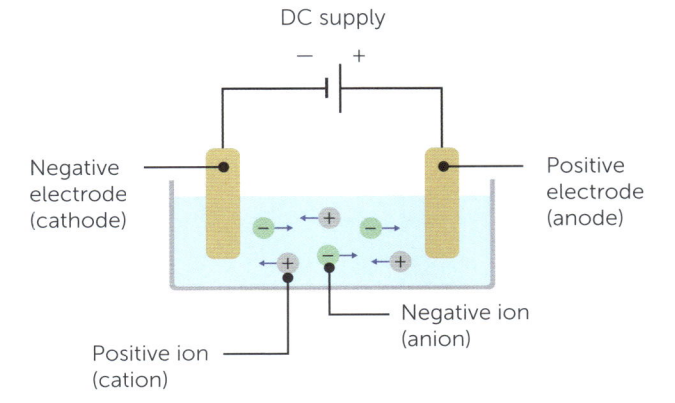

1. Give the meaning of direct current, dc. [1]
2. Explain why graphite and copper can act as electrodes. [2]
3. Explain why table sugar does not conduct electricity when molten or in solution. [2]

---

1. It is electric current that flows in one direction only.[1]
2. They conduct electricity[1] because they contain delocalised electrons[1].
3. Table sugar consists of small molecules[1] and does not have charged particles that are free to move[1].

Weak acids   Higher Tier only

Reduction and oxidation processes happen at the electrodes:
- at the cathode, positively charged ions gain electrons and are **reduced**.
- at the anode, negatively charged ions lose electrons and are **oxidised**.

You can represent these reactions using **half equations**. You can revise redox reactions on **page 62** and half equations on **page 4**.

AQA GCSE Chemistry 8462 / 8464 – Topic 4

ELECTROLYSIS OF MOLTEN IONIC COMPOUNDS

Molten ionic compounds decompose into elements.

Binary ionic compounds

A **binary** ionic compound consists of a metal element and a non-metal element. This means that potassium oxide (K_2O) is a binary ionic compound, but potassium hydroxide (KOH) is not.

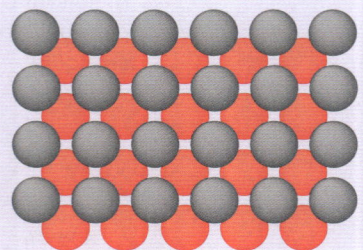

Potassium oxide structure

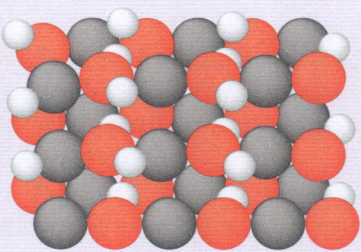

Potassium hydroxide structure

Binary ionic compounds decompose into the two elements they contain during electrolysis. The metal element forms at the negative charged electrode and the non-metal element forms at the positive electrode.

Half equations — Higher Tier only

At the cathode, potassium ions gain electrons and are reduced: $K^+ + e^- \rightarrow K$

At the anode, oxide ions lose electrons and are oxidised: $2O^{2-} \rightarrow O_2 + 4e^-$

---

1. A teacher passed an electric current through molten lead bromide using graphite electrodes. Orange-brown vapour was observed at the anode and silvery liquid at the cathode.
    (a) Explain the teacher's observations. [3]
    (b) Explain the function of the graphite electrodes. [2]
2. (a) Predict the product formed at each electrode during the electrolysis of molten zinc chloride. [2]
    (b) **Higher Tier:** Write half equations for the reactions that occur at the electrodes. [2]

    1. (a) Lead ions move to the cathode[1] and are discharged as lead atoms[1]. Bromide ions move to the anode[1] and are discharged as bromine, which escapes as a vapour[1].
       (b) They conduct the electric current into the molten lead bromide[1] but are inert because they do not become part of either product[1].
    2. (a) Zinc at the negative charged electrode[1], chlorine at the positive electrode[1].
       (b) $Zn^{2+} + 2e^- \rightarrow Zn$[1] $2Cl^- \rightarrow Cl_2 + 2e^-$[1]

EXTRACTING METALS USING ELECTROLYSIS

Aluminium is more reactive than carbon, so it is extracted using electrolysis.

Extracting metals

Metals that are less reactive than carbon can be extracted by reducing their ores with carbon.

Some metals are less reactive than carbon but cannot be extracted this way because they react with carbon. Tungsten could be extracted by reducing tungsten oxide with carbon, but tungsten carbide forms during the process.

Metals that are more reactive than carbon must be extracted using electrolysis. Large amounts of energy are used to produce the electric currents needed, so this process is usually more expensive than heating with carbon.

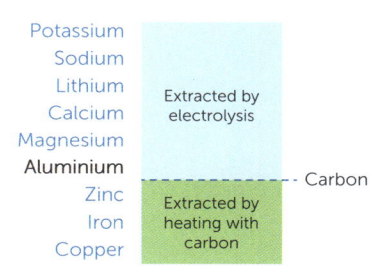

Manufacturing aluminium

Aluminium ore contains aluminium oxide, which is insoluble in water. It would be very expensive to produce molten aluminium oxide because its melting point is very high (around 2072 °C). To reduce the amount of energy needed, it is dissolved in molten **cryolite** (which melts at 950 °C).

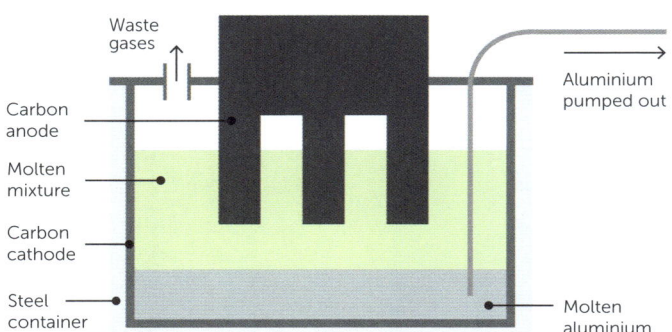

During the electrolysis of the molten mixture of aluminium oxide and cryolite, aluminium ions are discharged at the cathode and oxide ions are discharged at the anode.

Higher Tier only

At the cathode: $Al^{3+} + 3e^- \rightarrow Al$ **At the anode:** $2O^{2-} \rightarrow O_2 + 4e^-$

Explain why the carbon anodes must be replaced frequently during the manufacture of aluminium. [2]

Oxide ions are discharged as oxygen at the anodes. This reacts with the carbon in the anodes to produce carbon dioxide[1]. The anodes wear away as carbon escapes as carbon dioxide[1].

The huge electrical currents passing through the electrolysis cells in an aluminium smelter create powerful magnetic fields.

AQA GCSE **Chemistry 8462 / 8464 – Topic 4**

ELECTROLYSIS OF AQUEOUS SOLUTIONS

Ions from water

The electrolysis of aqueous solutions may produce gases at both electrodes. Some of the water molecules in water break down to produce hydrogen ions and hydroxide ions:

$$H_2O(l) \rightarrow H^+(aq) + OH^-(aq)$$

Depending on the other ions present, these ions may be discharged:
- H^+ at the cathode to make hydrogen
- OH^- at the anode to make oxygen.

Higher Tier only

An oxidation reaction happens at the anode:

$$4OH^- \rightarrow 2H_2O + O_2 + 4e^-$$

At the cathode

Hydrogen, H_2, is produced unless the metal is less reactive than hydrogen, when the metal is produced instead.

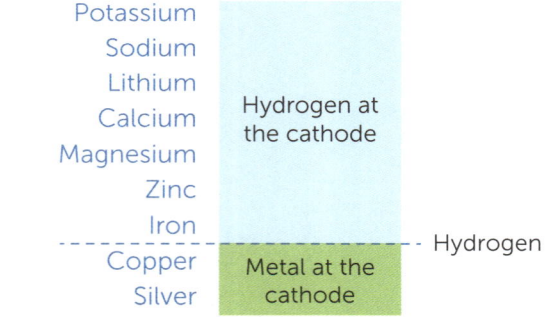

At the anode

Oxygen, O_2, is produced unless halide ions are present, when a halogen is produced instead:
- chlorine if chloride ions are present
- bromine if bromide ions are present
- iodine if iodide ions are present.

---

1. Predict the products formed at the cathode during the electrolysis of these solutions:
   (a) silver nitrate [1]
   (b) potassium bromide [1]
   (c) iron(III) chloride [1]

2. Predict the products formed at the anode during the electrolysis of these solutions:
   (a) zinc nitrate [1]
   (b) sodium carbonate [1]
   (c) zinc chloride [1]

3. Predict the products formed during the electrolysis of dilute sulfuric acid, $H_2SO_4(aq)$. [2]

1. (a) silver[1] (b) hydrogen[1]
   (c) hydrogen[1]
2. (a) oxygen[1] (b) oxygen[1]
   (c) chlorine[1]
3. Hydrogen at the negative electrode[1] and oxygen at the positive electrode[1].

Examples

Aqueous solution	Positive ions	At the cathode	Negative ions	At the anode
Sodium hydroxide	H^+ Na^+	Hydrogen	OH^-	Oxygen
Sodium chloride	H^+ Na^+	Hydrogen	OH^- Cl^-	Chlorine
Copper(II) chloride	H^+ Cu^{2+}	Copper	OH^- Cl^-	Chlorine
Copper(II) sulfate	H^+ Cu^{2+}	Copper	OH^- SO_4^{2-}	Oxygen

REQUIRED PRACTICAL 3 (9)
Investigating electrolysis

This required practical activity helps you develop your ability to draw, set up and use electrochemical cells, and to analyse the products formed.

Drawing electrochemical cells

A **scientific diagram** is a labelled representation of objects, not an artist's drawing. Make sure you:
- use a pencil and ruler
- draw straight lines where possible
- avoid unnecessary shading.

When you add a label to your drawing:
- draw a single straight line away from the object
- write a clear label at the end of the line.

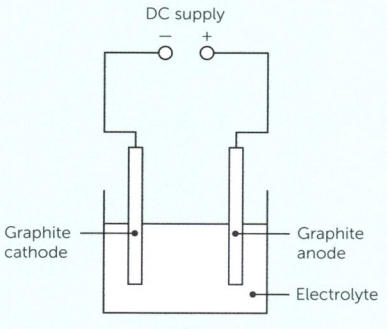

Gas tests

It is difficult to obtain enough hydrogen or oxygen for gas tests with simple electrochemical cells, but you should be able to test for the presence of **chlorine**. You can revise gas tests on **page 126**.

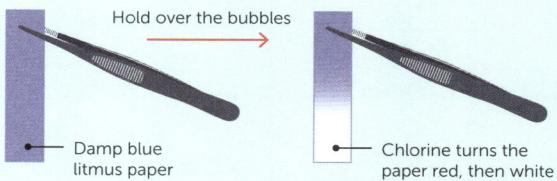

1. Predict what you would see during the electrolysis of copper(II) chloride solution. [2]
2. Explain why the graphite electrodes are described as inert in these experiments. [2]

   1. A brown-red coating on the negative electrode.[1] Bubbles coming off the positive electrode.[1]
   2. They provide a surface for the electrode reactions to happen on[1] but the carbon does not take part in the electrode reactions[1].

Electrolysis of aqueous solutions

The table shows the expected products of the electrolysis of four different **electrolytes** using graphite **electrodes**. You can revise how to predict these products on **page**s 74 and 76.

Electrolyte	Product at cathode (negative electrode)	Product at anode (positive electrode)
Copper(II) chloride solution	Copper	Chlorine
Copper(II) sulfate solution	Copper	Oxygen
Sodium chloride solution	Hydrogen	Chlorine
Sodium sulfate solution	Hydrogen	Oxygen

AQA GCSE Chemistry 8462 / 8464 – Topic 4

TOPIC 4

EXAMINATION PRACTICE

01 A student carried out an experiment to determine the relative reactivity of three metals (X, Y and Z). Mixtures of metals and metal oxides were heated, and the presence or absence of a reaction was recorded. The table shows the student's results.

	X oxide	Y oxide	Z oxide
Metal X		Reaction	No reaction
Metal Y	No reaction		No reaction
Metal Z	Reaction	Reaction	

01.1 Determine the order of reactivity, starting with the most reactive. [1]

01.2 Give a reason why the student did not heat metal X with X oxide. [1]

01.3 In a separate experiment, the student heated magnesium with copper(II) oxide. Magnesium oxide and copper were produced.
Explain which substance was oxidised and which substance was reduced. [2]

02 This question is about copper(II) nitrate.

02.1 Name the dilute acid needed to produce copper(II) nitrate. [1]

02.2 Give a reason why copper(II) nitrate cannot be made using copper. [1]

02.3 Name **two** insoluble solids that could be used to make copper(II) nitrate. [2]

03 A student added dilute hydrochloric acid to sodium hydroxide solution.

03.1 Compare the pH values of these two reactants before the experiment began. [1]

03.2 Write the balanced ionic equation for the neutralisation reaction that occurred. Include state symbols. [2]

04 Write the chemical formula of aluminium sulfide and sodium phosphate.
Formulae of ions: Al^{3+}, Na^+, S^{2-}, PO_4^{3-} [2]

05 Aluminium is extracted from aluminium oxide using electrolysis.

05.1 Give a reason why aluminium cannot be extracted by heating aluminium oxide with carbon. [1]

05.2 Explain why a molten mixture of aluminium oxide and cryolite is used in the process. [3]

05.3 Explain why the positive electrode must be replaced frequently. [3]

06 A teacher demonstrated electrolysis using molten zinc chloride and carbon electrodes.

06.1 Explain why molten zinc chloride conducts electricity. [2]

06.2 Name the products given off at each electrode. [2]

06.3 The teacher repeated the experiment using zinc chloride solution instead.
Compare the products with the products obtained from molten zinc chloride. [2]

Higher Tier only

07 Copper(II) chloride solution, $CuCl_2$(aq), is electrolysed using carbon electrodes.
Balance these half equations for the reactions that take place at each electrode.

07.1 Cu^{2+} + ... e^- → Cu [1]

07.2 ...... Cl^- → Cl_2 + ... e^- [1]

08 Magnesium reacts with dilute hydrochloric acid:

$$Mg + 2HCl \rightarrow MgCl_2 + H_2$$

08.1 Identify the substance that is oxidised in the reaction.
Explain your answer with the aid of a suitable half equation. [2]

08.2 Give a reason why this reaction is an example of a redox reaction. [1]

09 This question is about weak and strong acids.

09.1 Carbonic acid, H_2CO_3, is a weak acid. Explain what is meant by a weak acid. [2]

09.2 1.0×10^{-2} mol/dm³ sulfuric acid has a pH of 2. Sulfuric acid, H_2SO_4, is a strong acid.
Determine the pH of 1.0×10^{-5} mol/dm³ sulfuric acid. [1]

Chemistry only

10 A student carried out a titration to determine the concentration of a sample of dilute nitric acid.
The table shows the student's results.

	Titration 1	Titration 2	Titration 3	Titration 4
Volume of sulfuric acid in cm³	24.15	23.95	24.25	23.80

10.1 Concordant results are within 0.10 cm³ of each other.
Use the student's concordant results to calculate the mean volume of acid used. [2]

10.2 Explain why universal indicator solution is **not** a suitable indicator to use in acid-alkali titrations. [2]

Chemistry Higher Tier only

11 Sulfuric acid reacts with sodium hydroxide solution:

$$H_2SO_4 + 2NaOH \rightarrow Na_2SO_4 + 2H_2O$$

A student titrated 25.0 cm³ portions of 0.200 mol/dm³ sodium hydroxide solution with dilute sulfuric acid. The mean titre of sulfuric acid was 14.20 cm³.
Calculate the concentration of the sulfuric acid. [3]

4.5.1.1 **RPA4** **5.5.1.1** **RPA10**

EXOTHERMIC AND ENDOTHERMIC REACTIONS

Reactions can be **exothermic** or **endothermic**, depending on whether they transfer energy to or from the surroundings.

Energy changes in reactions

The amount of energy in the universe stays the same during a chemical reaction. This means that energy is conserved – the amount of energy transferred to or from the surroundings during a reaction is equal to the amount of energy lost or gained by the products.

Exothermic reactions

Exothermic reactions include:
- Combustion
- Many oxidation reactions
- Neutralisation

Energy is transferred **to** the surroundings in an exothermic reaction. The temperature of the surroundings increases if energy is transferred by heating.

Endothermic reactions

Endothermic reactions include:
- Thermal decomposition
- Citric acid and sodium hydrogen carbonate reacting together

Energy is transferred **from** the surroundings in an endothermic reaction. The temperature of the surroundings decreases if energy is transferred by heating.

Practical uses

Some cold packs used to treat sports injuries use endothermic reactions. Exothermic reactions are used in handwarmers for cold days. A pouch contains iron powder, and salty water held in carbon. Air enters when the pouch is opened. Oxygen reacts with the iron powder, and the salty water speeds up the reaction. As the iron oxidises, energy is transferred to the surroundings by heating.

---

1. Suggest the type of reaction (exothermic or endothermic) that is needed in a self-heating food can. Give a reason for your answer. [1]
2. Energy transfers occur when salts dissolve in water. Plan an experiment to distinguish between a salt that produces an exothermic change when it dissolves, and a salt that produces an endothermic change when it dissolves. [6]

1. Exothermic, because an exothermic reaction transfers energy to the surroundings and increases their temperature.[1]

2. Add the same volume of water to two polystyrene cups.[1] Measure and record the temperature in each cup.[1] Add a different salt to each cup and stir to dissolve.[1] Measure and record the new temperature in each cup.[1] The exothermic change causes the temperature to go up.[1] The endothermic change causes the temperature to go down.[1]

REQUIRED PRACTICAL 4 (10)
Exothermic and endothermic reactions

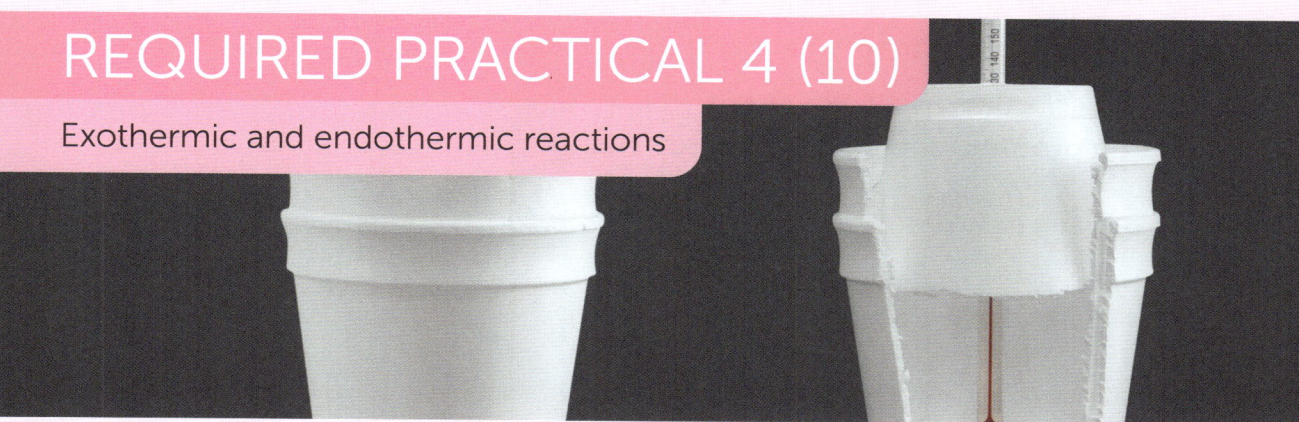

This required practical activity helps you develop your ability to carry out and monitor chemical reactions, including using substances carefully and safely.

Simple calorimeters

Reactions in solution cause changes in temperature. A **calorimeter** is used in experiments where these temperature changes are measured. Energy transfer between the surroundings and the reaction mixture is the greatest source of error in these experiments.

A simple calorimeter consists of a polystyrene cup held securely inside a beaker. The polystyrene and the air gap reduce energy transfers by **conduction**.

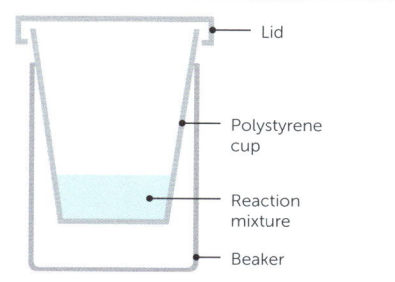

An example experiment

A typical experiment involves:
- recording the starting temperature of a solution of an acid, alkali or salt
- mixing the reactants
- recording the temperature of the reaction mixture over several minutes
- determining the maximum change in temperature.

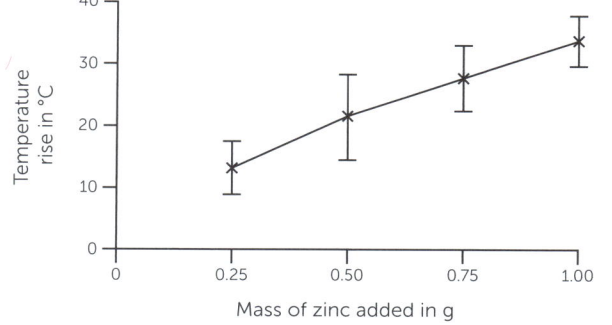

The graph shows the results of an investigation in which different masses of zinc powder were added to copper(II) sulfate solution. The reactions at each mass were repeated several times:
- each × shows a mean rise in temperature
- each I-shaped bar is an **error bar** that shows the **uncertainty** in the results.

---

1. Give a reason why there is a lid on the polystyrene cup in the diagram. [1]
2. Explain why a line graph was used, rather than a bar chart. [2]

   1. The lid reduces energy transfers by convection.[1]
   2. The temperature rise and mass of zinc added can both be measured.[1] They are continuous variables[1], so a line graph is suitable. Bar charts are suitable if a variable is categoric.[1]

4.5.1.2 5.5.1.2

REACTION PROFILES

Reaction profiles represent the energy changes during a chemical reaction.

Colliding particles

For a chemical reaction to happen:
- reactant particles must collide with one another, and
- they must have enough energy.

The activation energy is the minimum amount of energy needed for a reaction to happen.

Reaction profiles are also called energy level diagrams. They show the relative amounts of energy involved in reactions.

Exothermic reactions

In the reaction profile for an exothermic reaction:
- the energy level of the reactants is higher than the energy level of the products.

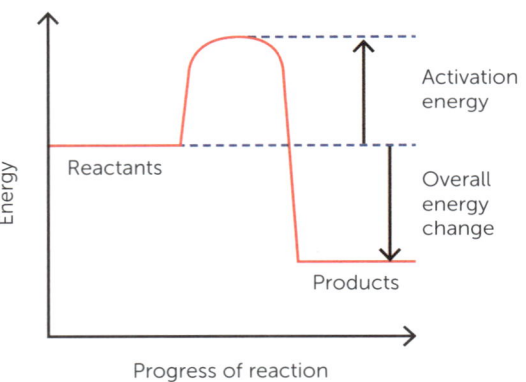

Endothermic reactions

In the reaction profile for an endothermic reaction:
- the energy level of the products is higher than the energy level of the reactants.

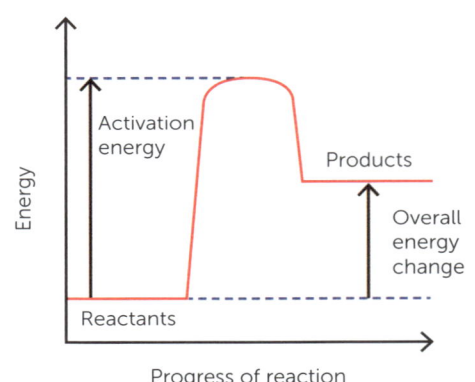

Breaking and making bonds   Higher Tier only

In a chemical reaction, energy is transferred:
- to the reactants in order to break bonds
- to the surroundings when bonds form in the products.

The overall energy change is the difference between the amounts of energy involved.

Compare the activation energy with the overall energy change in exothermic and endothermic reactions. [2]

The activation energy is always positive[1], but the overall energy change is negative in exothermic reactions and positive in endothermic reactions[1].

| 4.5.1.3 | 5.5.1.3 | Higher Tier |

THE ENERGY CHANGE OF REACTIONS

Bond energies

Bond energies can be used to calculate the overall energy change of a reaction. A **bond energy** is the amount of energy needed to break one mole of a given type of covalent bond. Different bonds have different bond energies, so chemists use tables of data rather than remembering their values. The table below shows the bond energies of some common bonds.

Bond	C–C	C–H	H–H	O–H	O=O	C=C	C=O
Energy in kJ/mol	347	413	436	464	498	612	805

As a chemical reaction happens, energy is:
- supplied to break bonds in the reactants – this is an endothermic process
- released when bonds form in the products – this is an exothermic process.

overall energy change of a reaction = (total energy supplied) − (total energy released)

	Exothermic	Endothermic
Amounts of energy involved	(energy in) < (energy out)	(energy in) > (energy out)
Sign of the overall energy change	Negative	Positive

1. Hydrogen reacts with oxygen to produce water: 2(H–H) + O=O → 2(H–O–H)
   (a) Calculate the overall energy change of this reaction. [4]
   (b) Explain what your answer to part (a) shows about the reaction. [2]
2. Propane undergoes a cracking reaction to produce ethene and methane:
   (a) Calculate the overall energy change of this reaction. [4]

$$\text{H-C(H)(H)-C(H)(H)-C(H)(H)-H} \rightarrow \text{H}_2\text{C=CH}_2 + \text{H-CH}_2\text{-H}$$

   (b) Explain what your answer to part (a) shows about the reaction. [2]

1. (a) Energy in to break bonds = (2 × 436) + 498 = 1370.[1]
   Energy out when bonds form = (4 × 464) = 1856.[1]
   Overall energy change = 1370 − 1856[1] = −486 kJ/mol.[1]
   (b) The reaction is exothermic[1] because the overall energy change is negative[1].
2. (a) Energy in to break bonds = (2 × 347) + (8 × 413) = 3998.[1]
   Energy out when bonds form = 612 + (4 × 413) + (4 × 413) = 3916.[1]
   Overall energy change = 3998 − 3916[1] = +82 kJ/mol.[1]
   (b) The reaction is endothermic[1] because the overall energy change is positive[1].

⭐ The eight C–H bonds in question 2 are unchanged. You may leave out unchanged bonds from your calculations if you are confident.

CHEMICAL CELLS AND FUEL CELLS

4.5.2.1–2 Chemistry

Chemical cells and fuel cells use exothermic reactions to produce electricity.

Chemical cells

Chemical cells need two **electrodes** and an **electrolyte**. You can make a simple chemical cell using two different metals dipped in a salt solution. A voltage (potential difference) develops between the two metals.

An electric **current** flows if the metals are connected in a complete circuit. Electrons flow in the circuit from the more reactive metal to the less reactive metal.

In general, the further apart the two metals are in the **reactivity series**, the greater the voltage.

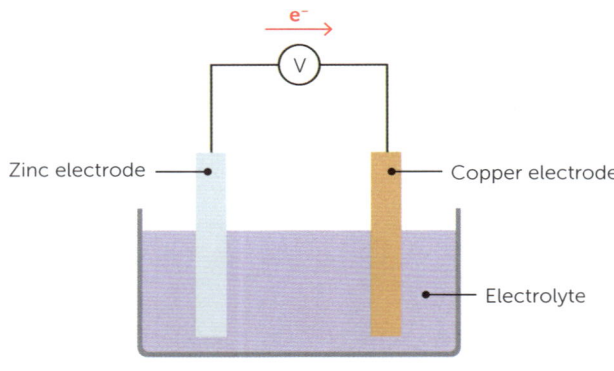

1. A student made chemical cells using four different metals (A, B, C and D) but the same electrolyte and right-hand electrode. The table shows the potential differences obtained.

Left-hand electrode	Right-hand electrode	Potential difference in V
Metal A	Copper	1.10
Metal B	Copper	2.71
Metal C	Copper	0.00
Metal D	Copper	0.59

Which metal was the most reactive?

Tick **one** box. [1]

Metal A ☐
Metal B ☐
Metal C ☐
Metal D ☐

1. Metal B.[1]

Non-rechargeable and rechargeable cells

Batteries consist of multiple cells connected in series to increase the voltage. The familiar alkaline batteries consist of **non-rechargeable** cells. They 'go flat' and fail to develop a voltage when one of their reactants is used up. The chemical reactions in **rechargeable** cells can be reversed by supplying an electric current.

Fuel cells

A **fuel cell** produces a voltage for a long time as it is supplied with a fuel and oxygen or air. Hydrogen fuel cells use hydrogen as their fuel. They may be used instead of chemical cells for some purposes. Hydrogen is oxidised to water in a hydrogen fuel cell:

$$2H_2 + O_2 \rightarrow 2H_2O$$

Although the overall reaction is the same as burning hydrogen, the reaction is an **electrochemical** one – it is achieved by the transfer of ions and electrons rather than by a flame.

2. Give **two** ways in which the voltage of chemical cells may be increased. [2]
3. Write **two** half equations for the electrode reactions in a hydrogen-oxygen fuel cell. [4]
4. Suggest **two** advantages of hydrogen fuel cells compared to rechargeable batteries for powering electric cars. [2]

   2. Use electrodes made from metals that are further apart in the reactivity series[1]. Connect two or more cells together in series[1] to make a battery.
   3. At the anode: H_2 [1] $\rightarrow 2H^+ + 2e^-$ [1]   At the cathode: $4H^+ + 4e^- + O_2$ [1] $\rightarrow 2H_2O$ [1]
   4. Electricity is supplied as long as hydrogen and air are supplied but the rechargeable battery will stop working when one of the reactants runs out[1]. The car's hydrogen fuel tank can be refilled more quickly than a battery can be recharged[1].

Lithium-ion batteries contain a lithium alloy electrode and a graphite electrode. They develop a voltage of around 3.8 V. Lithium and carbon are far apart in the reactivity series.

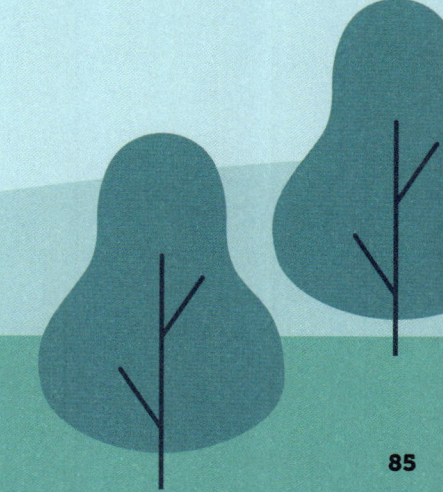

TOPIC 5

EXAMINATION PRACTICE

01 Describe what is meant by an exothermic reaction. [2]

02 A student added some water to a polystyrene cup, then recorded its temperature. The student dissolved some ammonium nitrate in the water, then recorded the temperature of the solution formed. The table shows the results.

Temperature at start in °C	Temperature at end in °C
19.4	16.7

02.1 Explain what the results show about the process of dissolving ammonium nitrate. [2]

02.2 Suggest **one** practical use of the observed changes. [1]

03 The diagram is a reaction profile for the reaction:

$$CaCO_3 \rightarrow CaO + CO_2.$$

It is not drawn to scale.

Explain what this reaction profile represents.

In your answer, you should refer to energy transfers and chemical bonds. [6]

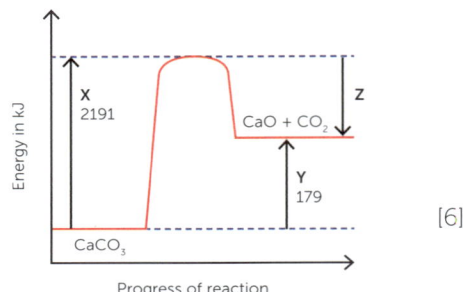

Higher Tier only

04 Hydrogen reacts with iodine to form hydrogen iodide:

$$H-H + I-I \rightarrow 2(H-I)$$

Bond energies in kJ/mol: I–I = 151, H–I = 298. The overall energy change is –9 kJ/mol.
Calculate the bond energy for the H–H bond. [4]

Chemistry only

05 Lithium-ion batteries are rechargeable but alkaline batteries are non-rechargeable.

05.1 Explain this difference between the two types of battery. [2]

05.2 Give **one** reason why both types of battery eventually stop working. [1]

05.3 A student makes a cell using a piece of zinc, a piece of copper, and copper sulfate solution.
The potential difference between the two metals is 1.10 V.
Explain **one** change needed to obtain a higher voltage. [2]

Chemistry Higher Tier only

06 Methanol fuel cells use methanol as a fuel instead of hydrogen.

06.1 Explain why air must be supplied to a methanol fuel cell. [2]

06.2 Balance the half equation for the anode reaction in the methanol fuel cell. [1]

$$CH_3OH + H_2O \rightarrow \ldots H^+ + \ldots e^- + CO_2$$

TOPICS FOR PAPER 2

Information about Paper 2:

Separate Chemistry 8462:

Written exam: 1 hour 45 minutes
Foundation and Higher Tier
100 marks
50% of the qualification grade
All questions are mandatory

Trilogy 8464:

Written exam: 1 hour 15 minutes
Foundation and Higher Tier
70 marks
16.7% of the qualification grade
All questions are mandatory

Specification coverage

The content for this assessment will be drawn from Topics 6–10 (Topics 13–17 Trilogy): The rate and extent of chemical change; Organic chemistry; Chemical analysis; Chemistry of the atmosphere; and Using resources.

It may also draw on fundamental concepts and principles from Topics 1–3 (Sections 5.1–5.3 Trilogy): Atomic structure and the periodic table; Bonding, structure, and the properties of matter; and Quantitative chemistry.

Questions

A mix of multiple-choice, structured, closed short answer and open response questions. They may include calculations.

Questions assess knowledge, understanding and skills.

| CHEMISTRY | TRILOGY |
| 4.6.1.1 | 5.6.1.1 |

CALCULATING RATES OF REACTION

The **rate of reaction** is a measure of how quickly a reaction takes place.

Mean rate of reaction

The more quickly a reaction happens, the greater its rate of reaction. A reaction with a high rate finishes sooner than a reaction with a low rate.

You can calculate the **mean rate** by recording the loss of a reactant during the reaction:

$$\text{mean rate of reaction} = \frac{\text{quantity of reactant used}}{\text{time taken}}$$

You can also calculate the mean rate of reaction by recording the formation of a product:

$$\text{mean rate of reaction} = \frac{\text{quantity of product formed}}{\text{time taken}}$$

1. A student recorded the volume of gas produced in a reaction. The table shows the results.

Time in s	0	30	60	90	120	150	180	210	240
Volume of gas in cm^3	0	24	34	59	70	76	79	80	80

(a) Plot these results on the grid. Draw a line of best fit. [3]

(b) Calculate the mean rate of reaction. Give the units. [3]

(a) All points correct to within ± ½ a small square[2] (allow 1 mark if 6 or 7 are correct), best fit (must be smooth and ignore the anomalous result)[1].

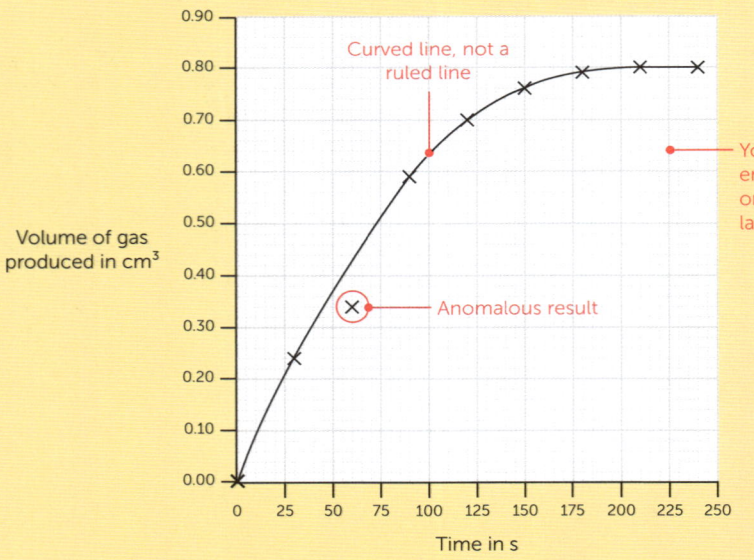

(b) Mean rate = $\frac{80 \text{ cm}^3}{210 \text{ s}}$ [1] = 0.38 [1] cm^3/s. [1]

In this example, no more gas was produced after 210 s. This means that the time taken for the calculation was 210 s, rather than 240 s. The student continued to record results after the reaction had finished.

Interpreting graphs

The **gradient** of the line on a rate of reaction graph gives the instantaneous rate.

Tangents and gradients

As the rate of reaction increases, the line on a rate of reaction graph becomes steeper and the gradient increases.

You can see this by drawing tangents to the curve. The line is horizontal when the reaction has stopped, and the gradient is zero.

If a rate of reaction is measured by recording changes in mass, its units are g/s.

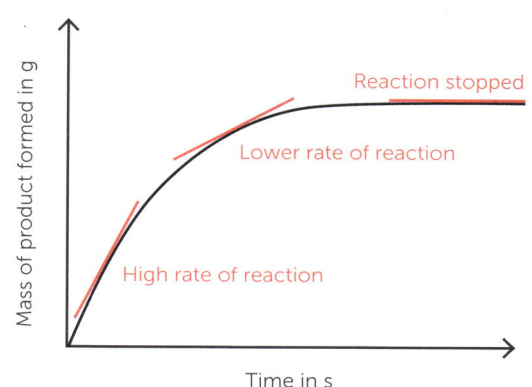

Rates from graphs **Higher Tier only**

You can find the rate of reaction at a given time during a reaction:
- Draw the tangent to the curve at that time.
- Calculate the gradient of the tangent.

When you draw your tangent, make sure that your ruler edge is the same distance either side of the curve at that time.

Draw two lines to your tangent so that you make a triangle as large as you can. This improves the accuracy of your gradient.

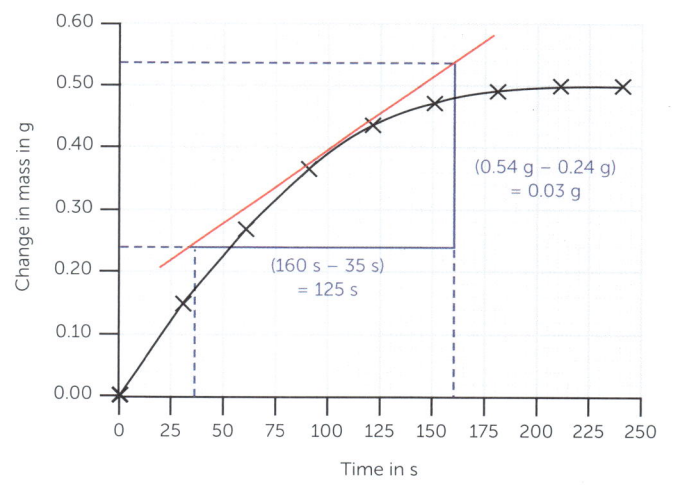

---

2. Describe what happens to the rate of reaction as the reaction proceeds. [2]

   Higher Tier only: In the second graph, calculate the rate of reaction at 100 s. Give the units. [3]

3. In a reaction, 0.10 mol of magnesium are used up in 50 s. Calculate the mean rate of reaction. Give the units. [2]

   2. The rate of reaction decreases as the reaction goes on[1] eventually becoming zero[1].

   rate of reaction = $\frac{0.54 \text{ g} - 0.24 \text{ g}}{160 \text{ s} - 35 \text{ s}}$ = $\frac{0.30 \text{ g}}{125 \text{ s}}$[1] = 0.0024[1] g/s[1]

   3. Mean rate = (0.10 mol)/(50 s) = 0.002[1] mol/s[1].

AQA GCSE Chemistry 8462 / 8464 – Topic 6

REQUIRED PRACTICAL 5 (11)

Measuring rate of reaction

This required practical activity helps you develop your ability to accurately make and record observations and measurements, to carry out and monitor chemical reactions, and to use substances carefully and safely.

Measuring volumes of gases

Fill the measuring cylinder with water, then turn it upside down under water in a trough.

As bubbles of gas go into the measuring cylinder, the water it contains is pushed downwards before you read the new volume.

You can use a **gas syringe** to measure the volume of any gas more accurately. They are made of glass and are expensive, so you may be given a measuring cylinder instead.

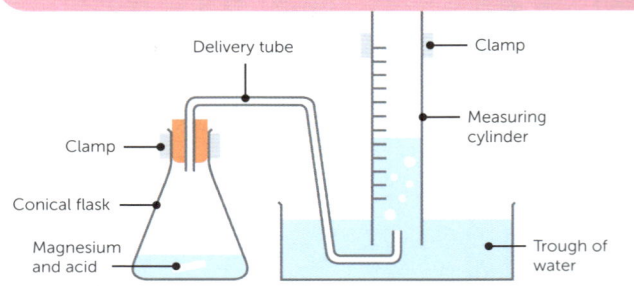

An example experiment

Magnesium reacts with dilute hydrochloric acid to produce magnesium chloride and hydrogen. You can use the rate of production of hydrogen to compare the rate of reaction at different concentrations of acid. The faster the gas syringe or measuring cylinder fills with gas, the greater the rate of reaction. You should control variables such as:

- the mass of magnesium – the greater the mass, the more gas is produced
- the size of the piece of magnesium – powders react more quickly than ribbon.

A table is the best way to record the results. For example:

Time in s	0	10	20	30	40	50	60	70	80	90
0.5 mol/dm³ acid										
1.0 mol/dm³ acid										

1. Explain why the temperature of the acid is a variable that should be controlled. [2]

   1. The rate of reaction depends upon the temperature of the reaction mixture.[1] If this was not kept the same each time, the rate of reaction would not only depend upon the concentration of the hydrochloric acid.[1]

Turbidity

The cloudier a liquid is, the greater its **turbidity**. Some reactions produce a **precipitate** that makes the reaction mixture more turbid. The greater the rate of reaction, the less time it takes for the reaction mixture to become so cloudy that you cannot see through it.

> You must carry out two different investigations for this practical activity. One like the experiment described on the opposite page, and another where you measure a change in colour or turbidity as the reaction carries on.

The disappearing cross investigation

Sodium thiosulfate solution reacts with dilute hydrochloric acid to produce four products, including sulfur. This is an insoluble yellow substance that makes the reaction mixture cloudy.

You can get a measure of the rate of reaction if you:
- place a conical flask containing the reaction mixture on top of a piece of paper with a cross drawn on it, then
- time how long it takes before you can no longer see the cross through the reaction mixture.

The total volume of the reaction mixture must be controlled, otherwise its depth will also affect the time taken. You can achieve this, and vary the concentration of sodium thiosulfate, by adding measured volumes of water.

The table shows some examples. A fixed volume of acid is needed to begin the reaction.

> You need to develop a **hypothesis** of your investigations – a proposal intended to explain observations or facts in advance of any experiments.

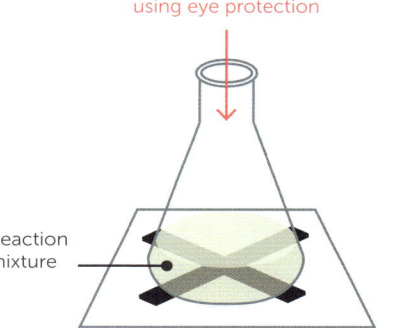

Look downwards using eye protection

Reaction mixture

Volume of sodium thiosulfate solution in cm³	10	20	30	40	50
Volume of water added in cm³	40	30	20	10	0
Concentration of sodium thiosulfate solution in g/dm³	8	16	24	32	40

2. Name **one** piece of apparatus that can be used to measure a change in colour. [1]

3. Refer to the disappearing cross investigation in this question.
   (a) Explain **one** precaution needed to work safely. [3]
   (b) Describe the expected results. [1]
   (c) Suggest **one** way in which the investigation could be extended. [1]

> 2. Colorimeter.[1]
>
> 3. (a) One of the products is sulfur dioxide gas[1], which is toxic and can cause breathing problems[1]. Avoid breathing in fumes when looking down at the reaction mixture.[1]
>
>   (b) As the concentration of sodium thiosulfate increases, the reaction time decreases.[1]
>
>   (c) Heat the sodium thiosulfate solution before adding the acid[1] so that the effect of temperature on the rate of reaction can be investigated[1].

AQA GCSE Chemistry 8462 / 8464 – Topic 6

FACTORS AFFECTING THE RATE OF REACTION

Five common factors that affect the rates of chemical reaction include the **concentrations** of reactants, the **presence of catalysts**, the **surface area** of solid reactants, the **pressure** or gases and the **temperature**.

Surface area

Explosions involve powdered substances that react together at very high rates.

As the surface area of a solid reactant increases, the rate of reaction increases. This is why powders react much more quickly than sheets and lumps.

Pressure

The **pressure** of a gas is a measure of how much force it exerts on its container walls and on objects inside it. If a chemical reaction involves a reactant in the gas state, the rate of reaction increases as the pressure increases.

Temperature

Cooking involves chemical reactions in food. It takes longer to cook eggs in hot water than it does to cook them in boiling water.

The **temperature** of a substance is a measure of the average **kinetic energy** of its particles. For a substance in a given state (solid, liquid or gas), heating causes its particles to move more quickly, and its temperature increases. As the temperature of a reaction mixture increases, the rate of reaction increases.

1. Describe how the rate of a reaction depends on the concentration of a reactant in solution. [1]
2. Describe what a catalyst is. [2]

   1. As the concentration of the dissolved reactant increases, the rate of reaction increases.[1]
   2. A catalyst is a substance that increases the rate of a reaction[1], but by the end of the reaction is unchanged chemically[1] or in mass[1].

You need to know about the effects of concentration, surface area, pressure, temperature, and catalysts.

COLLISION THEORY

CHEMISTRY 4.6.1.3 | **TRILOGY 5.6.1.3**

Collision theory is a scientific model that explains how different factors affect the rate of a reaction.

Colliding particles

A chemical reaction can only happen when:
- reactant particles collide with one another, **and**
- the colliding particles have enough energy.

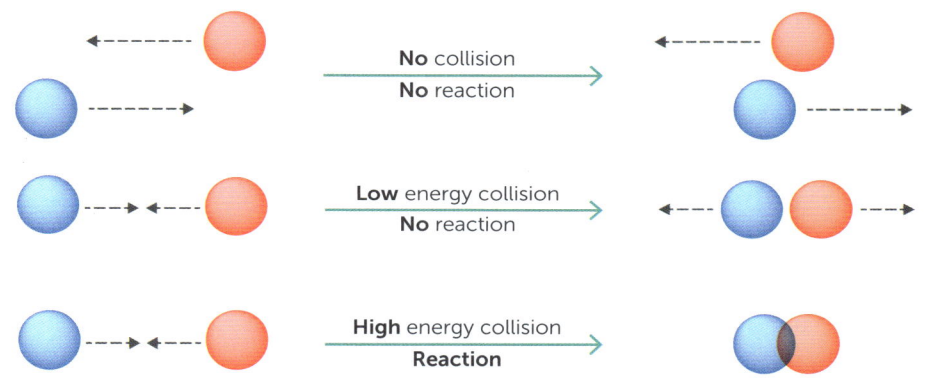

The **activation energy** is the minimum amount of energy that particles need for a reaction to happen. A collision that results in a reaction is a **successful collision**. The greater the **frequency** of successful collisions, the greater the rate of reaction. The table summarises how this works for increases in the concentration or pressure of a reacting substance.

Factor that increases	Frequency of collisions	Energy of particles	Frequency of successful collisions	Rate of reaction
Concentration	Increases	Does not change	Increases	Increases
Pressure	Increases	Does not change	Increases	Increases

1. For the same mass of a solid reactant, the rate of a reaction is greater if the reactant is a powder than if it is a lump. Explain this observation in terms of collision theory. [4]
2. A student explains the effect of the pressure of a reacting gas on reaction rate:
   "As the pressure increases, the number of collisions increases, so the rate of reaction increases."
   Explain why the student's explanation is **not** correct. [2]

   1. As the particle size decreases, the surface area to volume ratio increases.[1] More of the reacting solid is exposed to particles of the other reactant.[1] The energy of the particles does not change[1] but the frequency of successful collisions increases.[1]

   2. The student says 'number' of collisions instead of 'frequency' / 'rate' of collisions.[1] The number of collisions can become large over a long time, even in a slow reaction with a low rate of collisions.[1]

ACTIVATION ENERGY AND CATALYSTS

Effects of concentration

You might see two different effects of concentration on the rate of reaction:

- **Line 1** if the rate depends on the concentration of one of the reactants
- **Line 2** if the rate depends on the concentrations of two or more reactants

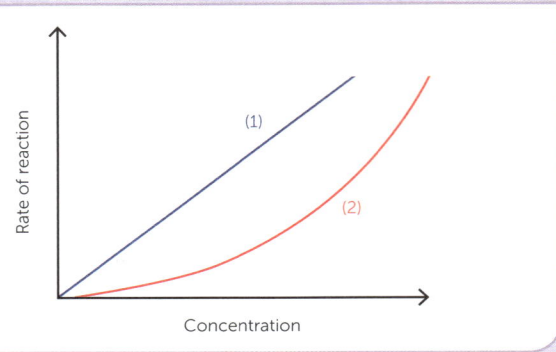

Effect of temperature

As the temperature of a reactant increases, the average energy of its particles increases. The percentage of particles that have the **activation energy** or more is much greater at high temperatures than it is at low temperatures.

Unlike increases in concentration, pressure or surface area, an increase in temperature causes:

- an increase in the frequency of collisions, **and**
- an increase in the energy of reactant particles.

These two factors combine to produce very large increases in the frequency of successful collisions, and in the rates of reactions. You are likely to observe reactions starting, or happening much faster, when the reaction mixture is warmed up.

This graph is here just to help you understand what is happening. You do not need to remember it.

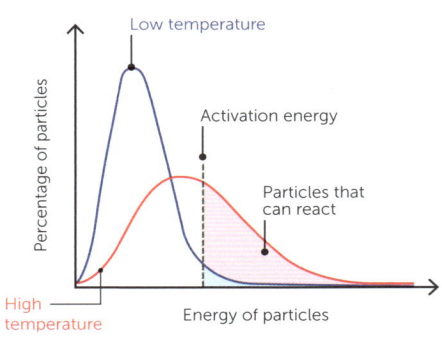

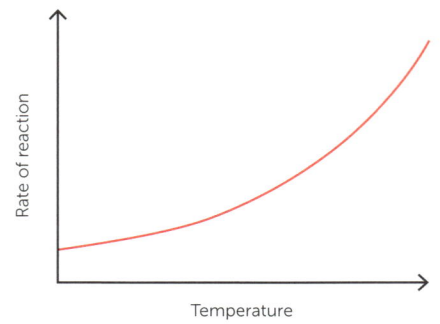

---

1. Explain what line 1 on the first graph shows. [3]
2. Predict how the rate of a reaction will depend on the surface area of a solid reactant. Give a reason for your answer. [2]

1. The graph shows that the rate of reaction is directly proportional to the concentration of a reactant.[1] This is because the line is straight, has a positive gradient[1] and passes through the origin[1].

2. The rate of reaction will be directly proportional to surface area.[1] This is because the number of reactant particles exposed at the surface increases as the surface area increases.[1]

Catalysts

A catalyst changes the rate of a chemical reaction without being used up in the reaction. Different reactions are catalysed by different catalysts.

You can tell if a substance in a reaction mixture acts as a catalyst because:

- it increases the rate of reaction
- it is **not** shown in the chemical equation for the reaction.

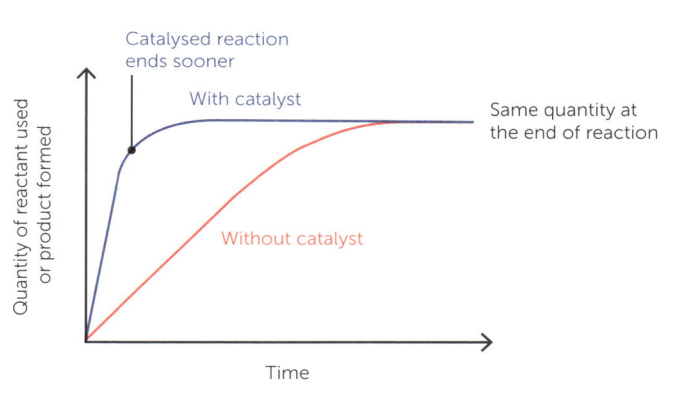

Activation energy

A catalyst provides a pathway that has a **lower activation energy** than the uncatalysed reaction. This can be shown on a **reaction profile** (you can revise these diagrams on **page 82**).

At a given temperature, a greater percentage of reactant particles will have the activation energy or more. The frequency of collisions will be the same as in the uncatalysed reaction, but a greater percentage of these collisions will be successful.

In the middle diagram on the opposite page, the effect is as if the activation energy line is moved to the left, letting more particles react.

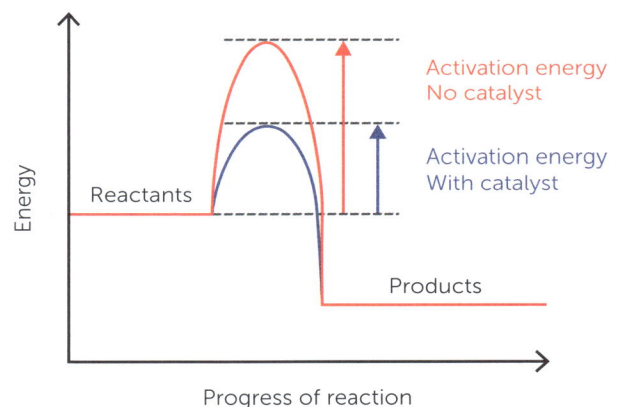

Cars are fitted with catalytic converters which contain precious metals. Hot exhaust gases heat these up and increase the rate of reaction to change harmful gases into CO_2 and water vapour.

1. A mixture of hydrogen peroxide solution and manganese(IV) oxide powder forms water and oxygen:

$$2H_2O_2(aq) \rightarrow 2H_2O(l) + O_2(g)$$

   Identify the catalyst in this reaction. Give a reason for your answer. [2]

2. (a) Describe what is meant by an enzyme. [2]
   (b) Give **one** example of the use of an enzyme in chemistry. [2]

---

1. Manganese(IV) oxide is the catalyst[1] because it does not appear in the balanced equation as a reactant or as a product[1].

2. (a) A protein that acts as a catalyst[1] in living things and other biological systems[1].
   (b) Fermentation to make ethanol[1] using enzymes in yeast[1].

REVERSIBLE REACTIONS

Some reactions are **reversible** and may not go to completion.

Representing a reversible reaction

In a reversible reaction, there is:
- a forward reaction – the reactants react together to produce products
- a reverse reaction – the products react together to produce the original reactants.

A split arrow symbol is used in equations for reversible reactions. In general:

$$A + B \rightleftharpoons C + D$$

Forward reaction: $A + B \rightarrow C + D$
Reverse reaction: $C + D \rightarrow A + B$

Ammonia reacts with hydrogen chloride to form ammonium chloride. The reaction is reversible:

$$NH_3(g) + HCl(g) \rightleftharpoons NH_4Cl(s)$$

White clouds of ammonium chloride form in the reaction. The photo shows how the reaction also happens with ammonia solution and hydrochloric acid.

A reaction between ammonia solution and hydrochloric acid

The reverse reaction is sometimes called the backward reaction.

Changing the reaction conditions

You can change the direction of a reversible reaction if you change the reaction conditions:
- change the temperature of the reaction mixture
- change the concentration of a reacting substance
- change the pressure of a reacting gas that is involved.

The reaction between anhydrous copper(II) sulfate and water is reversible:

anhydrous copper(II) sulfate (white) + water ⇌ hydrated copper(II) sulfate (blue)

1. When heated in a test tube, solid ammonium chloride decomposes to form ammonia and hydrogen chloride.
   Explain what you expect to see where the test tube is cold. [2]
2. The reaction between anhydrous copper(II) sulfate and water is exothermic.
   Describe the reverse reaction. [2]

   1. Solid ammonium chloride will form on the cold sides of the test tube[1] because heating makes the reaction go in one direction, and cooling makes it go in the opposite direction[1].
   2. It is endothermic[1] and transfers the same amount of energy as the forward reaction[1].

4.6.2.3 5.6.2.3

EQUILIBRIUM

Reversible reactions can reach **equilibrium** in closed systems.

Using up reactants

In any chemical reaction, the amounts of reactants gradually decreases as the reaction carries on. This means that the **rate of reaction** also decreases.

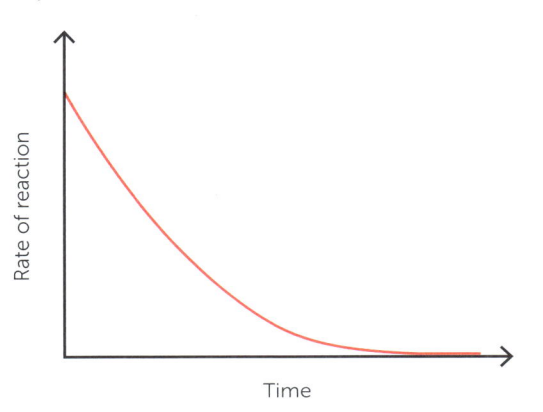

1. Magnesium reacts with dilute hydrochloric acid to produce magnesium chloride solution and hydrogen.
   Give **two** reasons why this reaction cannot reach equilibrium in an open beaker. [2]

2. A student says that the concentrations of all the reacting substances in a reaction at equilibrium are equal. Explain why the student is incorrect. [2]

   1. The reaction is not reversible[1] and one of the reacting substances, hydrogen, can escape[1].

   2. The concentrations of each substance stay the same[1] but different from each other[1].

Reaching equilibrium

In a reversible reaction, the amounts of products gradually increases as the reaction carries on. This means that the rate of the reverse reaction also increases. In a closed system such as a test tube or conical flask with a stopper on, or a reacting solution in a beaker, the rates of the forward reaction and reverse reaction eventually become exactly the same.

The reaction is described as having reached equilibrium.

In a reversible reaction at equilibrium:
- the forward and reverse reactions still keep going, and at the same rate
- the relative amounts of all the reacting substances do not change.

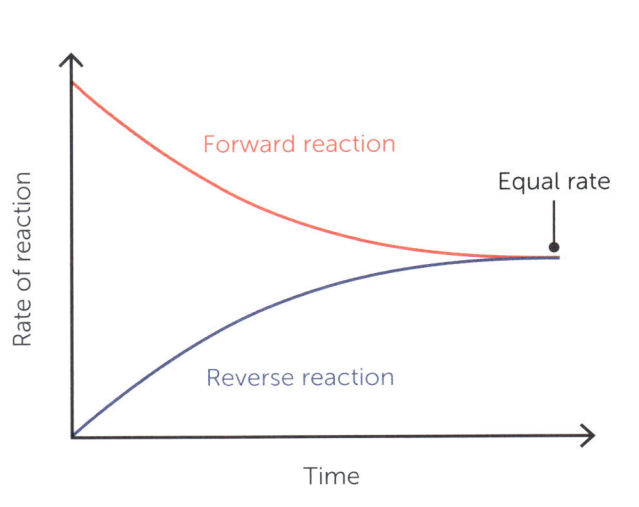

You might imagine equilibrium as like walking the wrong way on an escalator or travelator. You appear to stay in the same position if the rate of movement in each direction is the same.

AQA GCSE **Chemistry 8462 / 8464** – Topic 6

LE CHATELIER'S PRINCIPLE

You can use **Le Chatelier's principle** to predict how changing the conditions can affect a reaction at equilibrium.

Equilibrium position

The equilibrium position gives you an idea of the relative amounts of the reacting substances in a reaction at equilibrium. In the Haber process, nitrogen reacts with hydrogen to make ammonia:

$$N_2(g) + 3H_2(g) \rightleftharpoons 2NH_3(g)$$

The equilibrium position can lie to the left or to the right of the equation, depending on whether the relative amounts of reactants are greater than, or less than, the relative amounts of products.

$N_2(g) + 3H_2(g) \rightleftharpoons 2NH_3(g)$
Equilibrium position lies to the left
High relative amounts of reactants

$N_2(g) + 3H_2(g) \rightleftharpoons 2NH_3(g)$
Equilibrium position lies to the right
High relative amounts of products

A reaction at equilibrium responds to a change in reaction conditions, causing a movement in the equilibrium position. It moves to the left or to the right to counteract the change imposed.

Catalysts

In a reversible reaction, a **catalyst** increases the rates of the forward and reverse reactions by the same percentage, so:
- the equilibrium position stays the same, **but**
- equilibrium is reached more quickly.

This means that catalysts are still useful in industrial processes that involve reversible reactions, such as the Haber process.

An industrial ammonia plant

Predictions

Le Chatelier's principle does not explain why a change in a reaction condition leads to a change in the equilibrium position, but it does let you predict the outcome:
- you do not have to predict the actual amounts of each reacting substance, **but**
- you do have to predict whether the relative amounts of products will increase or decrease.

---

1. Describe Le Chatelier's principle. [2]
2. (a) Give **three** changes in conditions that can cause a change in equilibrium position. [3]
   (b) Give **one** change in conditions that cannot change the equilibrium position. [1]

1. It is the idea that when a change in conditions is imposed on a system at equilibrium[1], the system responds in a way that counteracts this change[1].
2. (a) Concentration[1], temperature[1], pressure[1].
   (b) Adding a catalyst.[1]

4.6.2.5 5.6.2.5 **Higher Tier**

CONCENTRATION CHANGES AND EQUILIBRIUM

You can also use **Le Chatelier's principle** to predict how changing the concentration of a reacting substance can affect the equilibrium position.

Rate of reaction

The rate of reaction increases as you increase the concentration of a reacting substance in solution.

For a reaction at equilibrium, if you add more of one of the reactants:
- the rate of the forward reaction increases
- more reactants are used up and more product forms
- the rates of the forward and reverse reactions gradually become equal again
- a new equilibrium position is reached, further to the right than before.

If you add more of one of the products instead, a new equilibrium position is reached again, but this time it is further to the left than before.

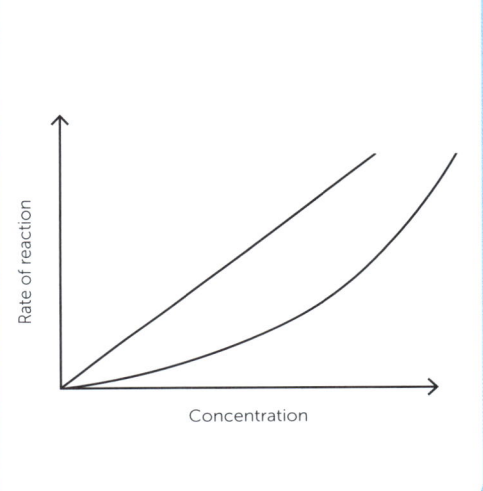

Making predictions

If you increase the concentration of a reacting substance, the equilibrium position moves in a direction away from this substance.

Think about the general equation for a reversible reaction:

$$A + B \rightleftharpoons C + D$$

If you increase the concentration of A or B, the position of equilibrium will move to the right. The effect of this is to use up some A and B, and to produce more C and D. In this way, the system responds to counteract the addition of extra reactants.

Remember that the reaction does not 'know' how to respond to changes in conditions.

Ethanol reacts with ethanoic acid to produce ethyl ethanoate and water. The reaction is reversible and takes place in aqueous solution. Sulfuric acid is a catalyst for this reaction.

Predict the effect on the equilibrium concentration of ethyl ethanoate of:

(a) increasing the concentration of ethanol. [1]

(b) decreasing the concentration of ethanoic acid. [1]

(c) decreasing the concentration of sulfuric acid. [1]

(a) It will increase.[1]

(b) It will decrease.[1]

(c) There will be no change.[1]

AQA GCSE **Chemistry** 8462 / 8464 – Topic 6

TEMPERATURE CHANGES AND EQUILIBRIUM

An increase in temperature shifts the equilibrium position in the direction of the **endothermic change**.

Exothermic and endothermic changes

If a reversible reaction is **endothermic** in one direction, it is **exothermic** in the opposite direction. The amount of energy transferred is the same in each direction.

Methanol can be manufactured from carbon monoxide and hydrogen:

$$CO(g) + 2H_2(g) \rightleftharpoons CH_3OH(g)$$

exothermic −91 kJ/mol

+91 kJ/mol endothermic

If the temperature of this reaction mixture is increased:
- the equilibrium position shifts to the left, because this is in the direction of the endothermic change
- the relative amount of methanol at equilibrium is decreased.

If you wanted to increase the relative amount of methanol at equilibrium, you would need to decrease the temperature of the reaction mixture. The equilibrium position would then shift to the right, in the direction of the exothermic change.

Steam reforming is used to manufacture hydrogen. It involves the reaction between methane and steam:

$$CH_4 + H_2O \rightleftharpoons CO + 3H_2$$

Energy change for forward reaction = −206 kJ/mol

Explain how the temperature of the reaction mixture may be changed to increase the equilibrium yield of hydrogen. [3]

Reduce the temperature[1] because this favours the exothermic reaction[1], and the forward reaction that produces hydrogen is exothermic[1].

Interpreting graphs

You can tell from a suitable graph whether a manufacturing process involves an exothermic reaction or an endothermic reaction. Ammonia is manufactured from nitrogen and hydrogen:

$$N_2 + 3H_2 \rightleftharpoons 2NH_3$$

The forward reaction, the one that produces ammonia, is exothermic. As the temperature is increased, the equilibrium moves to the left and the equilibrium yield of ammonia is decreased. A graph of equilibrium yield of product against temperature will have a negative gradient.

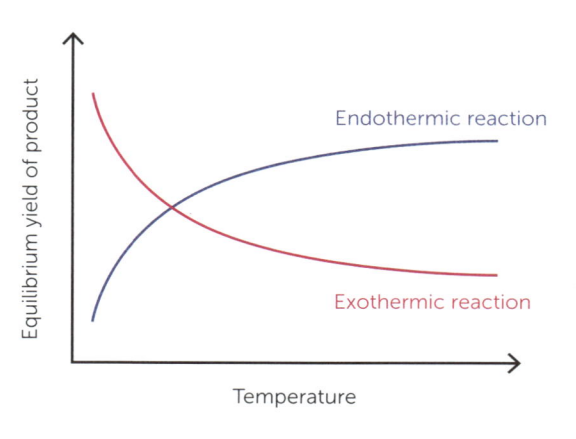

4.6.2.7 **5.6.2.7** **Higher Tier**

PRESSURE CHANGES AND EQUILIBRIUM

You must look at the balanced equation to predict the effect of changing the pressure of a reversible reaction.

Making predictions

If you increase the pressure of a reacting gas, the equilibrium position moves towards the side of the balanced equation that has the smallest amount of gas molecules.

For example, sulfur dioxide reacts with oxygen to form sulfur trioxide:

$$2SO_2(g) + O_2(g) \rightleftharpoons 2SO_3(g)$$

Reactant side: $2 + 1 = 3$ gas molecules. Product side: 2 gas molecules.

Changing the pressure has no effect on the equilibrium position if there are no reacting gases.

There are fewer molecules of gas on the right-hand side. If the pressure is increased:
- the equilibrium position shifts to the right
- the relative amount of sulfur trioxide at equilibrium increases.

If you reduce the pressure instead, the equilibrium position shifts to the left, and the relative amounts of sulfur dioxide and oxygen at equilibrium increase.

Summary

Change in conditions	Effect on equilibrium position
Increase in concentration	Shifts away from the substance that is increased in concentration
Increase in temperature	Shifts in the direction of the endothermic change
Increase in pressure	Shifts towards the side of the equation with fewest gas molecules
Catalyst added	No change

The time taken to reach equilibrium is decreased in each situation shown in the table.

1. Calcium carbonate decomposes when heated:
   $$CaCO_3(s) \rightleftharpoons CaO(s) + CO_2(g)$$
   Explain why the yield of calcium oxide is reduced if the pressure is increased. [2]

2. Hydrogen reacts with iodine:
   $$H_2(g) + I_2(g) \rightleftharpoons 2HI(g)$$
   Explain the effect of decreasing the pressure on the equilibrium position. [2]

   1. The equilibrium position moves to the left[1] because there are molecules of gas on the product side of the equation but no molecules of gas on the reactant side[1].
   2. No change[1] because the number of gas molecules is the same on each side of the equation[1].

AQA GCSE **Chemistry 8462 / 8464 — Topic 6**

TOPIC 6

EXAMINATION PRACTICE

01 A lump of calcium carbonate reacts with dilute hydrochloric acid. Give **one** change that could be made to the lump to increase the rate of reaction. [1]

02 A student reacted 0.08 g of calcium with excess dilute hydrochloric acid.
The graph shows how the volume of hydrogen given off changed during the reaction. Line C represents the student's results.

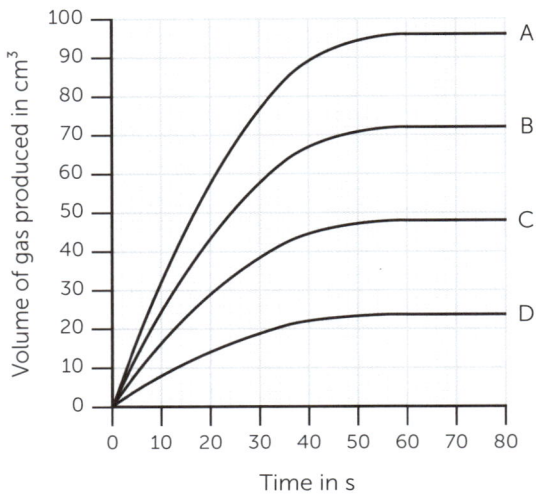

02.1 Describe how the student could measure the volume of hydrogen given off. [1]
02.2 Calculate the mean rate of reaction for the results shown by line C. Give the units. [3]
02.3 The student repeated the experiment using 0.12 g of calcium. Identify the line on the graph that represents the results from this experiment. Give a reason for your answer. [3]
02.4 **Higher Tier only:** Determine the rate of reaction at 40 s shown in line A. [3]

03 Give the meaning of the term activation energy. [1]

04 Sodium thiosulfate solution reacts with dilute hydrochloric acid:

$$Na_2S_2O_3(aq) + 2HCl(aq) \rightarrow 2NaCl(aq) + H_2O(l) + SO_2(g) + S(s)$$

04.1 Explain why the rate of this reaction decreases when water is added to the reaction mixture. [2]
04.2 Explain why the rate of this reaction increases when the temperature is increased. [3]

05 One of the stages in the manufacture of sulfuric acid involves producing sulfur trioxide from a mixture of sulfur dioxide, oxygen and vanadium(V) oxide:

$$2SO_2 + O_2 \rightleftharpoons 2SO_3$$

05.1 Identify the catalyst in this reaction. Give a reason for your answer. [2]
05.2 Explain how a catalyst works. [2]

06 Cobalt chloride is a substance used to test for the presence of water:

$$\text{anhydrous cobalt chloride} + \text{water} \rightleftharpoons \text{hydrated cobalt chloride}$$
$$\text{(Blue)} \qquad\qquad\qquad\qquad \text{(Pink)}$$

06.1 Give the meaning of the ⇌ symbol. [1]
06.2 Predict what you would see if anhydrous cobalt chloride became damp. [1]
06.3 Suggest a way to produce anhydrous cobalt chloride from hydrated cobalt chloride. [1]
06.4 The forward reaction is exothermic. Describe **two** features of the reverse reaction that can be deduced from this statement. [2]

07 A reversible reaction may reach equilibrium.
07.1 Give a reason why a reversible reaction in a stoppered boiling tube may reach equilibrium, but the same reaction cannot reach equilibrium if the stopper is removed. [2]
07.2 The rate of the forward reaction in a certain reversible reaction at equilibrium is 0.50 g/s. Explain what the rate of the reverse reaction will be. [2]

08 Magnesium ribbon reacts with hydrochloric acid: $Mg(s) + 2HCl(aq) \rightarrow MgCl_2(aq) + H_2(g)$
Plan an investigation to show how the concentration of hydrochloric acid affects the rate of reaction. [6]

Higher Tier only

09 Converters change harmful waste gases from car engines into less harmful ones:

$$2NO_2(g) + 4CO(g) \rightleftharpoons N_2(g) + 4CO_2(g)$$

Give a reason why these converters need catalysts. [1]

10 Ethene reacts with chlorine to produce dichloroethane:

$$C_2H_4(g) + Cl_2(g) \rightleftharpoons C_2H_4Cl_2(g)$$

Energy change for forward reaction = −218 kJ/mol

10.1 Explain the effect of increasing the pressure on the equilibrium yield of dichloroethane. [2]
10.2 Explain the effect of increasing the temperature on the equilibrium yield of dichloroethane. [2]

11 Iron(III) ions and thiocyanate ions react together in aqueous solution:
The mixture quickly reaches equilibrium.
Explain what you expect to happen when iron(III) chloride solution is added. [3]

$$\text{iron(III) ion} + \text{thiocyanate ion} \rightleftharpoons \text{thiocyanatoiron(III) ion}$$
$$Fe^{3+}(aq) \qquad SCN^-(aq) \qquad\qquad FeSCN_2{}^+(aq)$$
$$\text{(Brown)} \qquad \text{(Colourless)} \qquad\qquad \text{(Deep red)}$$

CRUDE OIL, HYDROCARBONS AND ALKANES

4.7.1.1 **5.7.1.1**

Crude oil consists of many compounds, most of which are **hydrocarbons**.

Crude oil

Crude oil and **natural gas** formed over millions of years from once-living plants and animals (mainly plankton). They are **finite resources** because they take so long to form or may not be being formed anymore. They are being used faster than they are being replaced.

Millions of years →

Plankton and algae die and are buried in mud

Ancient remains are exposed to high pressures and temperatures

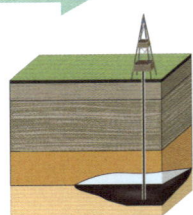

Crude oil and natural gas in rocks

---

1. Butane is an alkane. Its molecules contain four carbon atoms.

   1. (a) Determine the molecular formula of butane. [1]

   (b) Draw the displayed structural formula of butane. [1]

1. (a) Number of hydrogen atoms = (2 × 4) + 2 = 10, so the formula is C_4H_{10} [1]

(b)
```
    H H H H
    | | | |
H — C—C—C—C — H
    | | | |
    H H H H
```
[1]

Hydrocarbons and alkanes

Hydrocarbons are compounds of hydrogen and carbon only. There are several types of hydrocarbon, but the ones in crude oil are mostly **alkanes**.

The alkanes are a **homologous** series, a 'family' of compounds with similar chemical properties and the same general formula:

$$C_nH_{2n+2}$$

(where n stands for the number of carbon atoms in the molecule)

The table shows the first three alkanes.

Name of alkane	Methane	Ethane	Propane
Molecular formula	CH_4	C_2H_6	C_3H_8
Displayed structural formula	H—C(H)(H)—H	H—C(H)(H)—C(H)(H)—H	H—C(H)(H)—C(H)(H)—C(H)(H)—H

104 ClearRevise

FRACTIONAL DISTILLATION AND PETROCHEMICALS

The different hydrocarbons in crude oil are separated by **fractional distillation**.

Fractional distillation

Fractional distillation relies on the different **boiling points** of the different components in a mixture. You can revise fractional distillation of mixtures on **page 9**.

It involves:
- heating the mixture to **evaporate** its components
- cooling the vaporised components so they **condense** at different temperatures.

The fractional distillation of crude oil happens in a metal tower called a fractionating column.

The column gets cooler towards the top, so it has a **temperature gradient**. Different **fractions** have different ranges of boiling points, so they condense at different heights in the column.

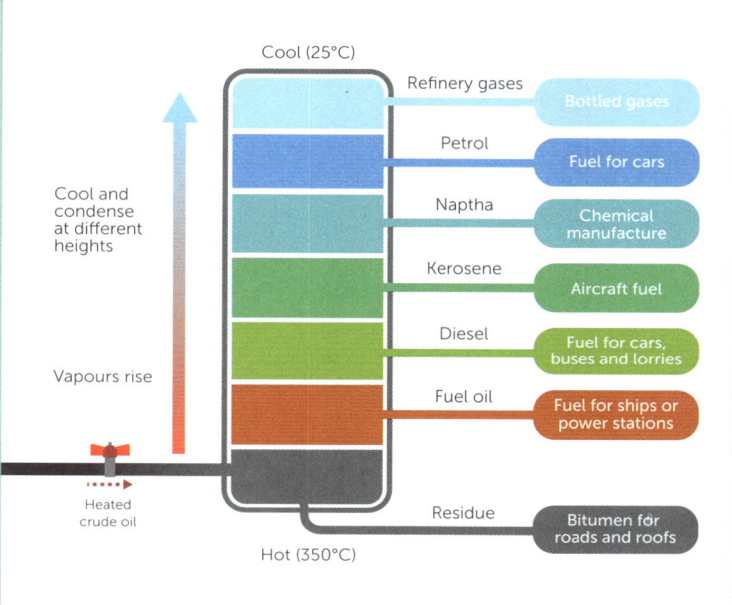

A substance occupies less volume when it is in the liquid state than when it is in the gas state. This makes LPG more suitable for transporting petroleum gases.

Fuels and petrochemicals

Crude oil provides many of the **fuels** that we use in everyday life. It also provides **feedstock** (starting materials) for the petrochemical industry. **Petrochemicals** are useful substances and materials produced from crude oil. They include detergents, lubricants, polymers (plastics) and solvents.

2. Place the following fractions in decreasing order of boiling point: diesel oil, heavy fuel oil, kerosene, LPG (liquefied petroleum gases), petrol. [1]
3. Give a reason why a huge number of different natural and synthetic carbon compounds are possible. [1]

    2. Correct order: heavy fuel oil, diesel oil, kerosene, petrol, LPG. [1]
    3. Carbon atoms can form homologous series or families of similar compounds. [1]

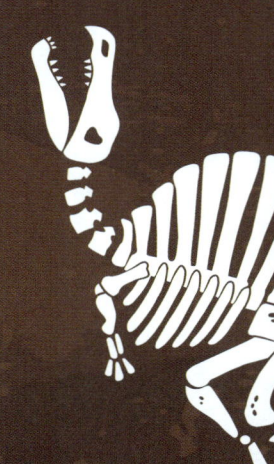

PROPERTIES OF HYDROCARBONS

Some **physical properties** of hydrocarbons depend on the size of their molecules.

Boiling point, viscosity and flammability

As the size of a hydrocarbon's molecules increases:
- the boiling point and viscosity increase
- the flammability decreases.

The **viscosity** of a substance is a measure of how easily it flows. A very viscous liquid has a high viscosity. It is 'thick' rather than runny, and does not flow easily.

These properties influence the use of hydrocarbons as fuels. For example, propane is in the gas state at room temperature, so it also has a very low viscosity. Propane ignites very easily. It is used in bottled gases for camping and for homes that are not connected to the mains gas supply.

Heavy fuel oil contains very large hydrocarbon molecules. It is a very viscous liquid that ignites with difficulty. This makes it suitable as a fuel for large ships, rather as a fuel for cars.

Getting more difficult to boil

Getting 'thicker' and less runny

Getting more difficult to ignite

Complete combustion

Combustion or burning is an **oxidation** reaction. It transfers energy to the surroundings, mainly by heating. Combustion in excess oxygen or air is called complete combustion. In general, for the complete combustion of a hydrocarbon:

hydrocarbon + oxygen → carbon dioxide + water

The hydrogen and carbon in the molecules gain oxygen and are oxidised.

Gases have very low viscosities, so they flow easily to fill their containers.

---

1. Write a balanced equation for the complete combustion of propane, C_3H_8. [2]
2. The hydrocarbon molecules in diesel oil are larger than the hydrocarbon molecules in petrol. Compare the boiling point, viscosity and flammability of these two fuels. [2]

   1. $C_3H_8 + 5O_2 \rightarrow 3CO_2 + 4H_2O$ 1 mark for correct formulae[1], 1 mark for correct balancing[1].
   2. Two from: Petrol has a lower boiling point[1], lower viscosity[1] and higher flammability[1] than diesel oil.

CRACKING

Cracking is a chemical process that breaks down hydrocarbons with large molecules to produce more useful substances with smaller molecules.

Breaking bonds

During cracking, **covalent bonds** are broken and new bonds form. A mixture of hydrocarbons with smaller molecules than the reactants is produced. For example:

$$C_6H_{14} \longrightarrow C_4H_{10} + C_2H_4$$

One of these products, ethene, is a type of hydrocarbon called an **alkene**. Unlike *alkane* molecules, alk*ene* molecules contain a carbon-carbon double bond, C=C.

Types of cracking

Steam cracking and catalytic cracking are two types of cracking. The table summarises the general conditions that these types of cracking need.

	Steam cracking	Catalytic cracking
Temperature in °C	750–900	450–550
Pressure in atmospheres	2–4	1
Catalyst	✗	✓

The catalysts in catalytic cracking are 'zeolites', substances that contain aluminium oxide.

1. Write a balanced equation for the cracking of octane, C_8H_{18}, to produce butane, C_4H_{10}, and one other product. [1]
2. Give **two** examples of how modern, everyday life depends on hydrocarbons. [2]

1. $C_8H_{18} \rightarrow C_4H_{10} + C_4H_8$ [1]
2. Hydrocarbons are used as fuels for cars[1], and for making polymers[1].

Usefulness of cracking

The petrochemical industry carries out cracking for two reasons:
- It helps to balance the supply of different hydrocarbons with the demand for them
- It produces alkenes, which are a feedstock for making **polymers**.

Hydrocarbons with large molecules are less useful as fuels than hydrocarbons with small molecules. Fractional distillation usually produces too much of the larger hydrocarbons and too little of the smaller hydrocarbons. Cracking helps to meet the high demand for fuels with smaller molecules.

The **supply** of something is how much can be made. The **demand** for something is a measure of how much people need or want.

ALKENES

Alkenes are hydrocarbons that contain carbon-carbon double bonds.

Making alkenes in the lab

Cracking can be carried out in the laboratory using liquid paraffin as the feedstock.

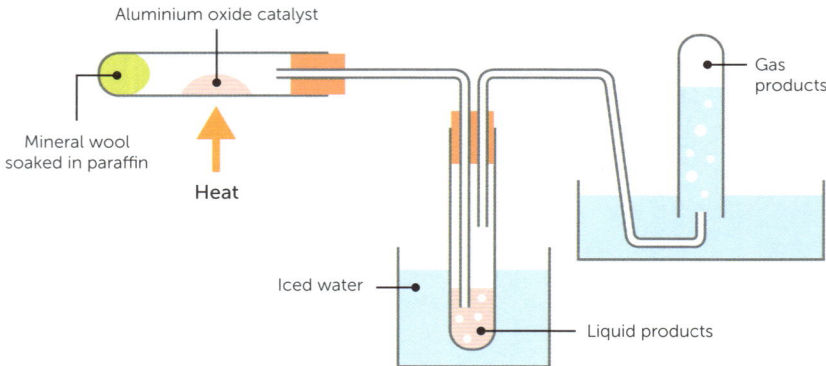

The catalyst must be heated very strongly before the paraffin is evaporated and passed over it. The iced water bath cools and condenses hydrocarbons with larger molecules. Hydrocarbons with smaller molecules carry on and are collected as gases.

Testing for alkenes

Alkene molecules contain C=C bonds. Alkanes do not. These C=C double bonds make alkenes more reactive than alkanes. This is the basis of a laboratory test for alkenes that uses **bromine water**.

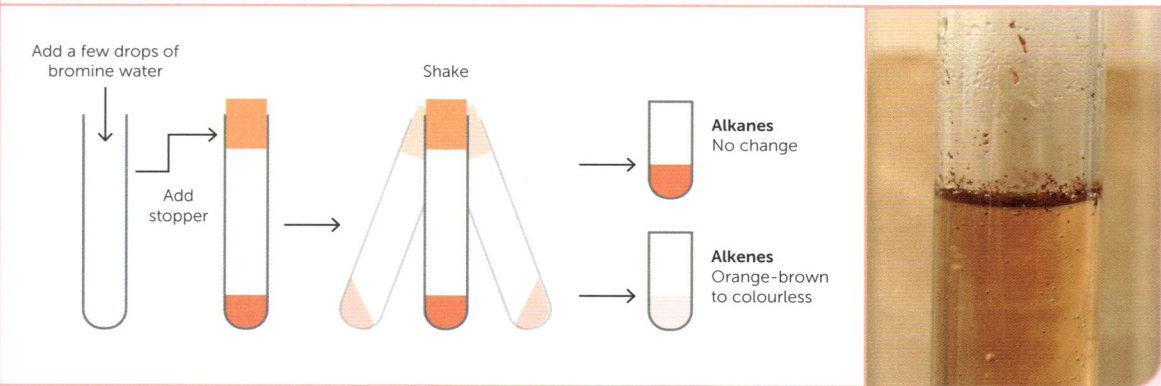

Liquid paraffin was cracked in the laboratory. Explain why the products of cracking decolourised bromine water but the liquid paraffin did not. [2]

Alkenes can decolourise bromine water, but it stays orange-brown when mixed with alkanes.[1]
The liquid paraffin did not contain alkenes but the products of cracking did.[1]

STRUCTURE AND FORMULAE OF ALKENES

Alkanes are **saturated** hydrocarbons. Alkenes are **unsaturated** hydrocarbons.

Homologous series

The alkenes are a **homologous series**, a 'family' of compounds with similar chemical properties and the same general formula:

$$C_nH_{2n}$$

(where n stands for the number of carbon atoms in the molecule)

The table shows the first three alkenes.

Name of alkene	Ethene	Propene	Butene
Molecular formula	C_2H_4	C_3H_6	C_4H_8
Displayed structural formula	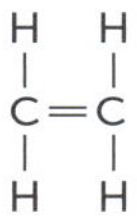 H H | | C=C | | H H	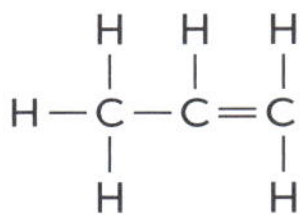 H H H | | | H—C—C=C | | H H	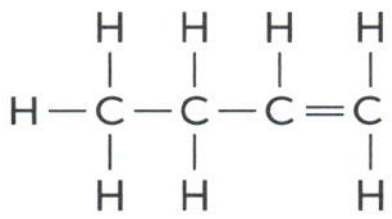 H H H H | | | | H—C—C—C=C | | | H H H

You might also see the displayed formula of ethene like this:

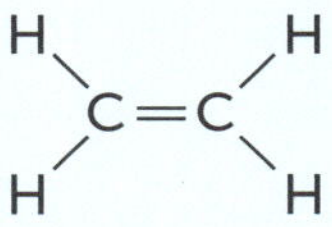

$$\begin{array}{cc} H & \quad H \\ \diagdown & \diagup \\ C & = C \\ \diagup & \diagdown \\ H & \quad H \end{array}$$

This matches the actual shape of an ethene molecule more closely.

Alternative structures

The C=C bond in alkenes with four or more carbon atoms may be located in more than one position. For example, it can be between the middle two carbon atoms in butene.

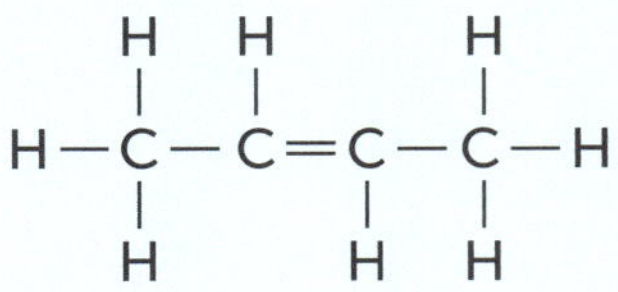

```
    H   H       H
    |   |       |
H—C — C = C — C—H
    |           |   |
    H           H   H
```

Molecules with the same molecular formula are called isomers. The two isomers of butene are called position isomers because the C=C is in different positions on the carbon chain.

---

1. Explain why pentene is unsaturated but pentane is saturated. [3]
2. Draw **two** possible displayed structural formulae for pentene. [2]

1. Pentene molecules contain a C=C bond but pentane molecules do not.[1] Pentene and pentane molecules both contain 5 carbon atoms[1] but pentene has two fewer hydrogen atoms[1].

2. 1 mark for each correct structure:

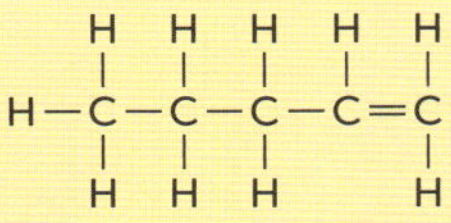

H H H H H [1]   H H   H [1]
\| \| \| \| \| \| \| \|
H—C—C—C—C=C   H—C— C=C—C—H
\| \| \| \| \| \|
H H H H   H   H

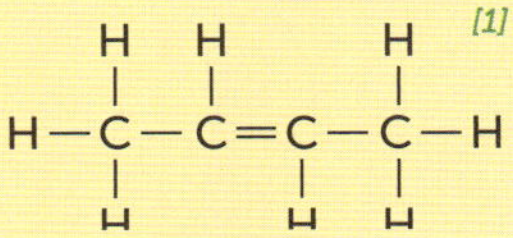

4.7.2.2 Chemistry

REACTIONS OF ALKENES

Unlike alkanes, alkenes often burn in air with smoky flames.

Organic compounds

Organic compounds consist of two or more elements, one of which must be carbon. This means that all hydrocarbons, including alkanes and alkenes, are organic compounds. Simple compounds that contain carbon are not classed as organic compounds. These include:
- carbon monoxide, CO
- carbon dioxide, CO_2
- carbonates such as calcium carbonate, $CaCO_3$.

The **functional group** of an organic compound is an atom, group of atoms or bonding that gives the substance its characteristic reactions. The functional group for alkenes is the C=C bond. This group is responsible for the way that alkenes can react with hydrogen, halogens and water. You can revise these reactions on **pages 110-112**.

Combustion

Alkenes undergo **complete combustion** in excess oxygen to form carbon dioxide and water:

   alkene + oxygen → carbon dioxide + water

However, **incomplete combustion** may also happen when alkenes burn in a limited supply of oxygen or in air. Some of the carbon atoms in the alkene molecules may not fully oxidise. When this happens, carbon particles (soot) and toxic carbon monoxide gas are produced:

   alkene + oxygen → carbon + carbon monoxide + water

The carbon particles make the flame smoky and they glow orange when hot. You can see the effects of incomplete combustion in the orange safety flame of a Bunsen burner (which is fuelled by alkanes not alkenes).

1. Name **one** type of reaction involving alkenes that does not depend on the C=C functional group. [1]
2. Hexene, C_6H_{12}, is a liquid alkene which is sometimes used as a laboratory solvent. Write a balanced equation for the incomplete combustion of hexene to produce water, and equal amounts of carbon and carbon monoxide. [2]
3. Suggest **two** reasons why alkenes are not used as fuels. [2]

   1. Combustion.[1]
   2. $C_6H_{12} + 4½O_2 \rightarrow 3C + 3CO + 6H_2O$ **or** $2C_6H_{12} + 9O_2 \rightarrow 6C + 6CO + 12H_2O$
      1 mark for correct formulae[1], 1 mark for balancing[1].
   3. Two points from: They burn with a smoky flame / incomplete combustion releases less energy than complete combustion[1]. They are important feedstock for the petrochemical industry[1]. Tendency for incomplete combustion means they are more dangerous given the carbon monoxide gas produced[1].

Addition reactions

Alkenes undergo **addition reactions** in which the C=C bond becomes a C–C bond.

Alkenes to alkanes

Alkenes react with hydrogen when heated in the presence of a nickel catalyst. For example:

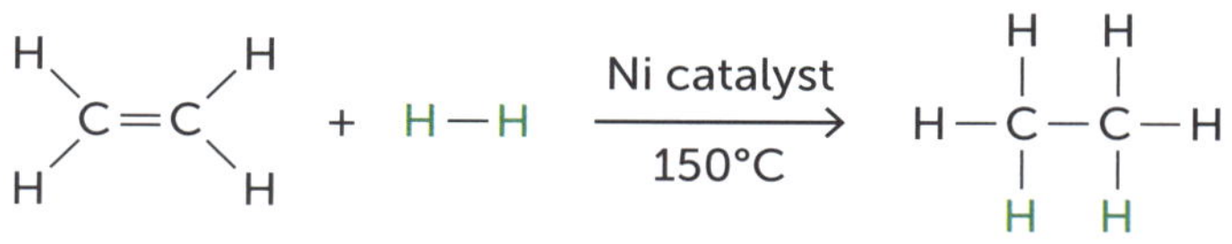

$$\begin{array}{c}H\\ \end{array}\!\!C\!=\!C\!\!\begin{array}{c}H\\ \end{array} + H\!-\!H \xrightarrow[150°C]{Ni\ catalyst} H-\overset{H}{\underset{H}{C}}-\overset{H}{\underset{H}{C}}-H$$

This reaction is called hydrogenation. It is an example of an addition reaction because two reactants add together to make a single product.

Alkenes to haloalkanes

Alkenes react with **halogens** at room temperature to produce **haloalkanes**.

Halogen	Type of compound formed	Speed of reaction
Chlorine	dichloroalkane	Fast
Bromine	dibromoalkane	Medium
Iodine	diiodoalkane	Slow

For example:

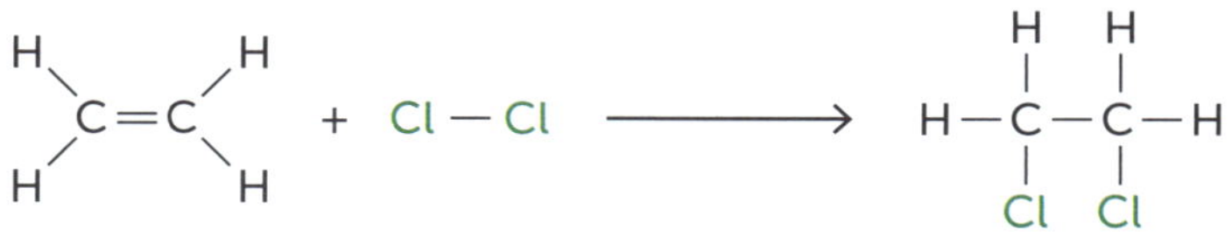

$$\begin{array}{c}H\\ \end{array}\!\!C\!=\!C\!\!\begin{array}{c}H\\ \end{array} + Cl\!-\!Cl \longrightarrow H-\overset{H}{\underset{Cl}{C}}-\overset{H}{\underset{Cl}{C}}-H$$

This type of reaction is called halogenation. It is also an example of an addition reaction.

---

4. Draw the displayed structural formulae of the following alkenes:

   (a) Propene [1]   (b) Butene [1]   (c) Pentene [1]

5. Draw the displayed structural formulae of the products of the following reactions:

   (a) Propene + chlorine [1]   (b) Butene + bromine [1]   (c) Pentene + iodine [1]

4. (a) H–C(H)(H)–C(H)–=C(H)(H) [1]
   (b) H–C(H)(H)–C(H)(H)–C(H)=C(H)(H) [1]
   (c) H–C(H)(H)–C(H)(H)–C(H)(H)–C(H)=C(H)(H) [1]

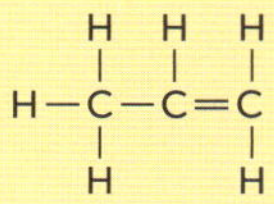

5. (a) H–C(H)(H)–C(H)(Cl)–C(H)(Cl)–H [1]
   (b) H–C(H)(H)–C(H)(H)–C(H)(Br)–C(H)(Br)–H [1]
   (c) H–C(H)(H)–C(H)(H)–C(H)(H)–C(H)(I)–C(H)(I)–H [1]

AQA GCSE Chemistry 8462 / 8464 – Topic 7

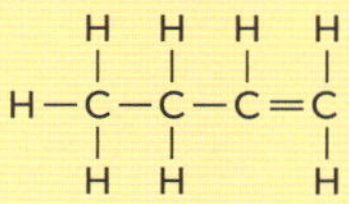

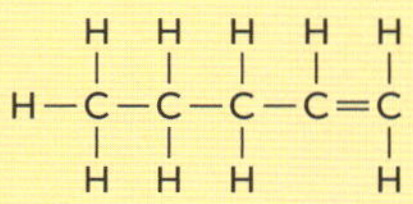

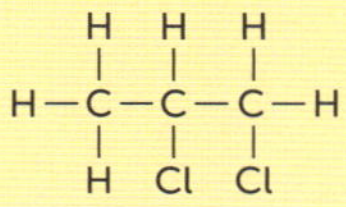

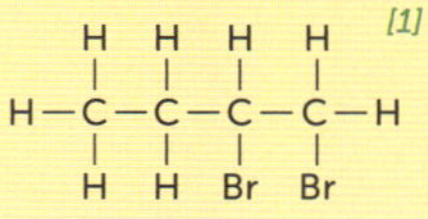

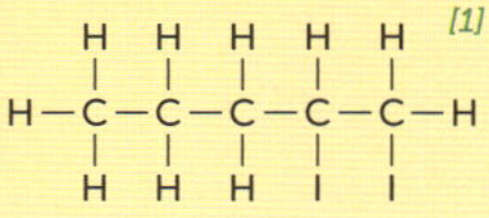

4.7.2.2 Chemistry

ALKENES AND WATER

Alkenes react with steam to produce **alcohols**.

Ethene and steam

The addition reaction between alkenes and water or steam is called **hydration**. Ethene reacts with steam to produce ethanol:

$$CH_2=CH_2 + H_2O \xrightarrow[\text{300°C 60 atmospheres}]{\text{Phosphoric acid catalyst}} CH_3-CH_2-OH$$

Industrial production of ethanol

The reaction of ethene with steam is used industrially to produce ethanol. The reaction gives you an opportunity to revise some other areas of chemistry.

The **atom economy** is 100% because ethanol is the only product. All the atoms in the reactants become atoms in the product. For atom economy, see **page 54**.

$$C_2H_4(g) + H_2O(g) \rightleftharpoons C_2H_5OH(g) \quad \text{energy change of the forward reaction = } -42 \text{ kJ/mol}$$

Higher Tier only

The high pressure shifts the **equilibrium position** to the right, which increases the equilibrium yield of ethanol. The high temperature is a compromise between a low yield of ethanol and a high rate of reaction. Increasing the temperature shifts the equilibrium position in the direction of the reverse reaction, but it ensures a high rate of reaction. You can revise **Le Chatelier's principle** on **pages 98–101**, and the choice of reaction conditions in industrial chemical processes on **page 159**.

Other alkenes

When other alkenes are involved, the –OH group is attached to a carbon atom away from the end of the molecule. Butene reacts with steam to form butanol. The –OH group is not attached to a carbon atom at the end.

$$H-\underset{H}{\overset{H}{C}}-\underset{H}{\overset{H}{C}}-\underset{O-H}{\overset{H}{C}}-\underset{H}{\overset{H}{C}}-H$$

---

1. Name the product formed when pentene reacts with steam. [1]
2. (a) Draw the displayed formula of propene, C_3H_6. [1]
   (b) Draw the displayed structural formula of propanol, which forms from propene. [1]

1. Pentanol[1].
2. (a) [1]

$$H-\underset{H}{\overset{H}{C}}-\underset{}{\overset{H}{C}}=\underset{H}{\overset{H}{C}}$$

(b) [1]

$$H-\underset{H}{\overset{H}{C}}-\underset{O-H}{\overset{H}{C}}-\underset{H}{\overset{H}{C}}-H$$

4.7.2.3 Chemistry

ALCOHOLS

The **alcohols** form a homologous series of organic compounds.

The –OH group

Alcohol molecules all have the functional group, –OH.

Name of alkene	Methanol	Ethanol	Propanol
Molecular formula	CH_3OH	CH_3CH_2OH	$CH_3CH_2CH_2OH$
Displayed structural formula	H–C(H)(H)–O–H	H–C(H)(H)–C(H)(H)–O–H	H–C(H)(H)–C(H)(H)–C(H)(H)–O–H

You should see that:
- The structural formulae show the atoms bonded to each individual carbon atom.
- The –OH group is written as OH in structural formulae, but drawn as –O–H.

The first three alcohols are fully **soluble** in water. They dissolve completely to form colourless solutions. Butanol is less soluble (only about 10 cm³ of it will dissolve in 100 cm³ of water).

Reactions of alcohols

Alcohols react with sodium to form **alkoxides** and hydrogen. For example:

ethanol + sodium → sodium ethoxide + hydrogen

Alcohols react with **oxidising agents** such as acidified potassium dichromate:

(alcohol in excess) ethanol + [oxidising agent] → ethanal + water
(oxidising agent in excess) ethanol + [oxidising agent] → ethanoic acid + water

The reaction mixture changes from orange to green during these reactions.

1. The structural formula of butanol is $CH_3(CH_2)_3OH$.
   (a) State **two** ways this information shows that butanol is an alcohol. [2]
   (b) Draw the fully displayed structural formula of butanol. [1]
2. Describe what you expect to observe when sodium is added to propanol. [2]
3. Name the organic product formed when methanol reacts with excess oxidising agent. [1]

   1. (a) Its name ends in 'ol'.[1] Its structural formula has the –OH functional group.[1]
      (b) H–C(H)(H)–C(H)(H)–C(H)(H)–C(H)(H)–O–H [1]
   2. Bubbling[1] (as hydrogen is given off) and the formation of an alkaline solution[1].
   3. Methanoic acid.[1]

AQA GCSE **Chemistry** 8462 / 8464 – Topic 7

4.7.2.3 Chemistry

FERMENTATION

Ethanol is produced by a biological process called **fermentation**.

Uses of ethanol and other alcohols

Alcohols are useful as fuels. Alcohols undergo complete combustion to produce carbon dioxide and water. For example, petrol contains ethanol:

$$CH_3CH_2OH + 3O_2 \rightarrow 2CO_2 + 3H_2O$$

Alcohols are used as solvents, such as propanol in inks. Ethanol has antiseptic properties. It is widely used in antibacterial and antiviral hand gels.

Manufacturing ethanol

Ethanol can be manufactured by the hydration of ethene (you can revise this process on **page 112**). However, most ethanol is produced by fermentation. This is a biological process that involves **enzymes** in microscopic fungi called **yeast**, and a sugar solution:

sugar $\xrightarrow{\text{yeast}}$ ethanol + carbon dioxide

The conditions needed are 20–35 °C and normal atmospheric pressure.

An aqueous solution of ethanol is produced. This is **filtered** to remove dead yeast and other insoluble materials. **Fractional distillation** is then used to separate ethanol from the other substances in the solution.

Ethanol is the alcohol found in beer, wine and other alcoholic drinks. Ethanol is produced on an industrial scale as a **biofuel**.

1. (a) Name **one** source of sugar for fermentation. [1]
   (b) Explain why it is important to exclude air from a fermentation mixture. [2]
   (c) Compare the conditions needed to produce ethanol by hydration of ethene and by fermentation of sugars. [3]

1. (a) Sugar beet / cane / wheat / barley.[1]
   (b) This prevents oxygen in air oxidising the ethanol[1] to form ethanoic acid[1].
   (c) Hydration uses a phosphoric acid catalyst but fermentation uses enzymes in yeast[1]. Hydration happens at high temperatures and pressures[1] but fermentation happens at moderate temperatures and normal pressure[1].

CARBOXYLIC ACIDS

The **carboxylic acids** form a homologous series of organic compounds.

The –COOH group

Carboxylic acid molecules all have the functional group, –COOH.

Name of carboxylic acid	Methanoic acid	Ethanoic acid	Propanoic acid
Structural formula	HCOOH	CH_3COOH	CH_3CH_2COOH
Displayed structural formula	H–C=O | O–H	H H–C–C=O | | H O–H	H H | | H–C–C–C=O | | | H H O–H

You should see that:
- the displayed structural formulae show the atoms bonded to each individual carbon atom
- the –COOH group is written as COOH in structural formulae, but drawn as –C=O
                                                                                                             \|
                                                                                                            O–H
- the –COOH group is attached to the end of the molecule.

Acidic properties

The first four carboxylic acids are fully **soluble** in water. They dissolve completely to form weakly acidic solutions. Vinegar is a dilute aqueous solution of ethanoic acid.

Just like other acids, carboxylic acids react with carbonates. For example, sodium carbonate reacts with ethanoic acid:

sodium carbonate + ethanoic acid →
sodium ethanoate + water + carbon dioxide

Higher Tier only

Carboxylic acids are weak acids because they only partially ionise in aqueous solution. The pH of carboxylic acids is higher than the pH of hydrochloric acid and other strong acids at the same concentration. See strong and weak acids on **page 72**.

2. The structural formula of butanoic acid is

$CH_3CH_2CH_2COOH$

(a) Describe **two** ways this information shows that butanoic acid is a carboxylic acid. [2]

(b) Draw the fully displayed structural formula of butanoic acid. [1]

3. Explain one observation seen when sodium carbonate is added to ethanoic acid. [3]

2. (a) Its name ends in 'anoic acid'.[1] Its structural formula has the –COOH functional group.[1]

(b) [1]
H H H
\| \| \|
H–C–C–C–C=O
\| \| \| \|
H H H O–H

3. Bubbling[1] because carbon dioxide gas is given off[1].

You can revise neutralisation reactions and salt formation on **page 64**.

REACTIONS OF CARBOXYLIC ACIDS

Carboxylic acids react with alcohols to form **esters** and water.

Esters

Esters have the functional group –COO–. The diagram shows the displayed structural formula of an ester called methyl butanoate.

```
      H   O   H   H   H
      |   ||  |   |   |
  H—C—O—C—C—C—C—H
      |       |   |   |
      H       H   H   H
```

Note that the –COO– group is usually shown 'backwards' in displayed structural formulae, as it is here.

Esters have 'fruity' smells. They are often used as food flavourings. For example:
- Methyl butanoate smells like pineapple.
- Ethyl ethanoate smells like pear drop sweets.

Ethyl ethanoate

Ethyl ethanoate is formed in the reaction between ethanol and ethanoic acid:

ethanol + ethanoic acid ⇌ ethyl ethanoate + water

A little hydrochloric acid may be added to the reaction mixture to act as a **catalyst**.

Ethyl ethanoate is used in glues and as a **solvent** in nail polish remover. The diagram shows its displayed structural formula. You do not need to know the name of any other ester.

```
      H   H       O   H
      |   |       ||  |
  H—C—C—O—C—C—H
      |   |           |
      H   H           H
```

> An ester is any organic compound that reacts with water to produce alcohols and acids. The reaction is reversible.

A mixture of propanol, methanoic acid and sulfuric acid reacts to form an ester and water:

propanol + methanoic acid ⇌ propyl methanoate + water

(a) Give a reason why sulfuric acid is included in the reaction mixture. [1]

(b) Describe what the symbol ⇌ in the word equation shows. [1]

(a) Sulfuric acid is added as a catalyst.[1]
(b) It shows that the reaction is reversible.[1]

ADDITION POLYMERISATION

Alkenes can react together to make **addition polymers**.

Monomers and polymers

Monomers are small molecules that join together to form very large molecules called **polymers**:

many monomer molecules → polymer molecule

Alkenes can take part in **addition reactions** because their molecules have a C=C group. For example, ethene molecules react together to form poly(ethene) molecules – 'poly' means 'many'. You can represent this reaction using displayed structural formulae and the idea of **repeating units**.

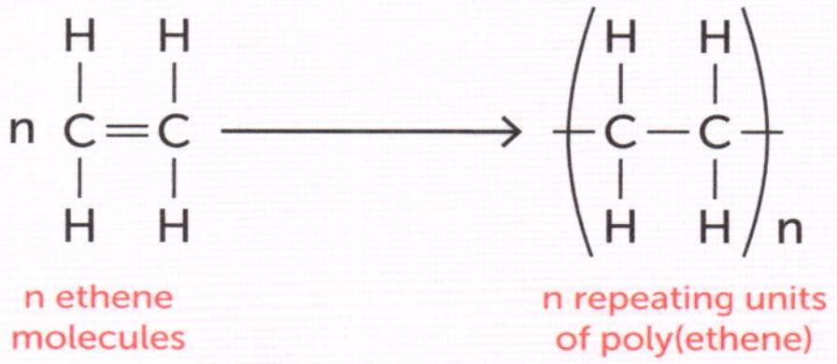

Notice that the repeating unit has the same number of atoms of each element as the monomer has. This is because the polymer is the only product of **addition polymerisation**.

Working out a repeating unit

You can work out the structure of a repeating unit when you are given the structure of the monomer. You could follow these steps to work out the repeating unit of poly(propene).

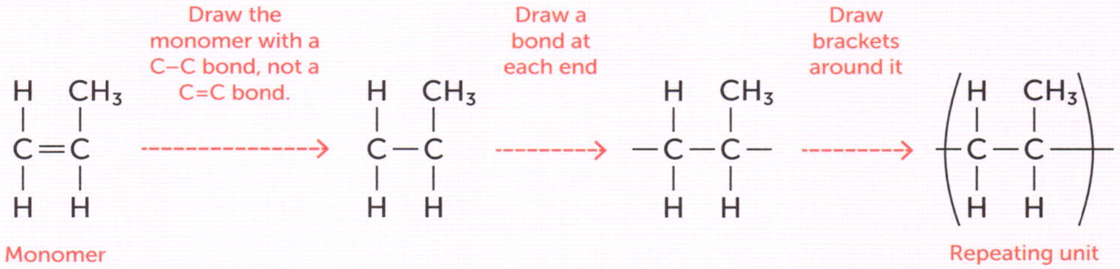

Once you have worked out the repeating unit, you can represent the polymerisation reaction.

The diagram shows a displayed formula of chloroethene.

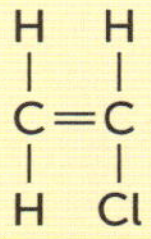

(a) Explain why chloroethene can act as a monomer. [2]
(b) Draw diagrams to represent the formation of poly(chloroethene). [3]

(a) Chloroethene contains a C=C bond[1] which lets it take part in addition reactions[1].
(b) In the repeating unit: C–C and bonds either side[1], remainder of structure correct[1], monomer, letters n and arrow drawn[1].

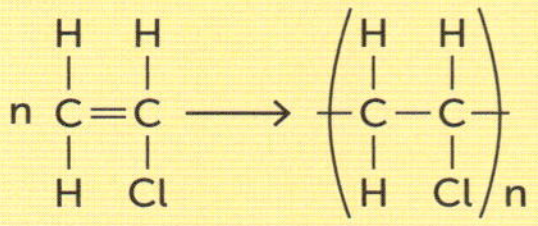

AQA GCSE Chemistry 8462 / 8464 – Topic 7

CONDENSATION POLYMERISATION

4.7.3.2 Chemistry — Higher Tier

Condensation polymers are made from monomers with two functional groups.

Monomers

Polyesters are condensation polymers formed from two different monomers:
- a diol – a molecule with two –OH groups
- a dicarboxylic acid – a molecule with two –COOH groups.

For example, ethanediol reacts with hexanedioic acid to make a larger molecule and a smaller molecule (commonly water and hence why these are called condensation reactions).

Ethanediol + Hexanedioic acid → Larger molecule + Smaller molecule

Diagrams like these are complicated, so they are often simplified by replacing the middle of each molecule by a rectangle.

$$HO-\square-OH \;+\; HOOC-\square-COOH \;\rightarrow\; HO-\square-O-CO-\square-COOH \;+\; H_2O$$

Diol + Dicarboxylic acid → Larger molecule (with ester group) + Water

Repeating units

Notice that the larger molecule produced in the reactions above has an –OH group at one end, and a –COOH group at the other end. They can react with these groups in other molecules to produce polymers. A polyester is a condensation polymer that contains many **ester groups**. You can follow these steps to work out the repeating unit of a polyester.

$$HO-\square-O-CO-\square-COOH \;\dashrightarrow\; O-\square-O-CO-\square-CO \;\dashrightarrow\; +\!(O-\square-O-CO-\square-CO)\!+$$

Larger molecule → Draw the ester without the H on OH or OH on COOH → Draw a bond at each end and brackets around it → Repeating unit

Once you have worked out the repeating unit, you can represent the polymerisation reaction.

Draw diagrams to represent the formation of a polyester from a diol and a dicarboxylic acid. [4]

$$n\,HO-\square-OH \;+\; n\,HOOC-\square-COOH$$
$$\downarrow$$
$$+\!(O-\square-O-CO-\square-CO)\!+_n \;+\; 2nH_2O$$

In the monomers: two –OH groups and two –COOH groups[1]. In the repeating unit: –O–CO– group and bonds and brackets either side[1], letters n and arrow drawn[1], 2n H_2O[1].

AMINO ACIDS AND DNA

DNA, complex carbohydrates and proteins are **naturally occurring polymers**.

Proteins and amino acids

Proteins are polymers made from **amino acid** monomers.

Higher Tier only

Amino acid molecules have two different functional groups, $-NH_2$ and $-COOH$. These groups allow amino acids to react together to form condensation polymers called **polypeptides**. Proteins are very large polypeptides that have a function in living organisms. Different polypeptides and proteins are made from different numbers of amino acids, joined together in different ways.

Higher Tier only: The diagram shows a structural formula of an amino acid called alanine. Draw diagrams to represent the formation of a polypeptide from alanine molecules. [3]

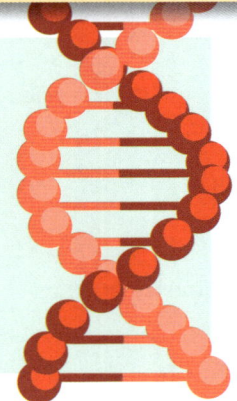

Correct structure repeating unit[1], monomer, letters n and arrow drawn[1], H_2O as the second product[1].

DNA

DNA (**deoxyribonucleic acid**) is the genetic material found in most cells. DNA molecules:
- are very large
- are made from four different monomers called **nucleotides**
- mostly consist of two polymer chains, wrapped around each other in a **double helix** structure.

The sequence of nucleotides in DNA forms a **genetic code**. Viruses and living organisms need this genetic code in order to develop and function correctly.

Complex carbohydrates

Carbohydrates are compounds of carbon, hydrogen and oxygen. Glucose is a carbohydrate. It is the monomer in complex carbohydrates such as starch and cellulose. Glucose monomers form chains and branches in these very large polymer molecules.

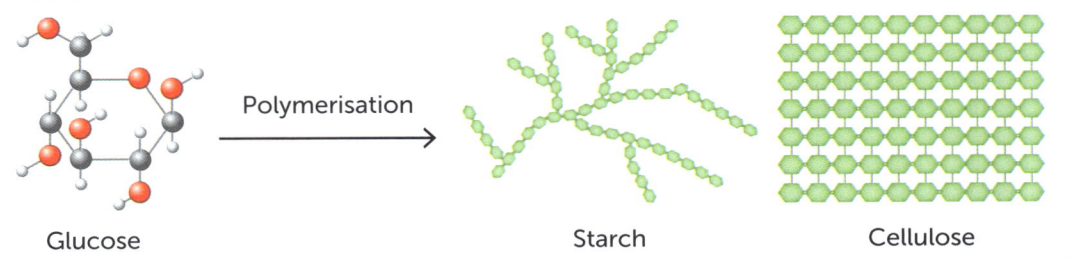

TOPIC 7

EXAMINATION PRACTICE

01 Decane, $C_{10}H_{22}$, is found in crude oil.
    01.1 Explain why decane is a hydrocarbon. [1]
    01.2 Explain, as fully as you can, why decane is an alkane. [2]

02 Draw the fully displayed formula of propane, C_3H_8. [2]

03 Crude oil is separated into useful fractions by fractional distillation.
    03.1 Name **two** crude oil fractions used as fuels. [2]
    03.2 Explain how fractional distillation of crude oil works. [4]

04 Hydrocarbons have different physical properties.
    04.1 Describe what is meant by the **viscosity** of a substance. [1]
    04.2 Describe how the boiling point and flammability of hydrocarbons change as their molecules become larger. [2]

05 Write a balanced equation for the complete combustion of pentane, C_5H_{12}. [2]

06 Hexane and hexene are liquid hydrocarbons. Hexane is an alkane and hexene is an alkene. Describe a chemical test to distinguish between these two liquids. [2]

07 Oil refineries crack hydrocarbon fractions from crude oil.
    07.1 Give **one** reason why cracking is carried out. [1]
    07.2 Compare the general reaction conditions in steam cracking and catalytic cracking. [2]
    07.3 Balance this equation for the cracking of hexadecane: [2]

$$\_\_\_\_\_ C_8H_{18} \rightarrow C_7H_{16} + C_3H_8 + \_\_\_\_\_ C_2H_4$$

Chemistry only:

08 Butene is an alkene. Butene molecules contain 4 carbon atoms.
    08.1 Predict the chemical formula of butene. [1]
    08.2 Give **one** reason why butene is described as an **unsaturated** hydrocarbon. [1]

09 Complete the displayed structural formula of propene. [1]

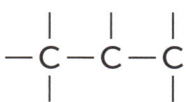

10 Give **one** reason why alkenes tend to burn in air with **smoky** flames. [1]

11 Ethene reacts with hydrogen and with chlorine.
Compare the conditions needed for these two reactions. [3]

12 Ethene, C_2H_4, reacts with steam to produce ethanol, CH_3CH_2OH.
  12.1 Complete the displayed structural formula of ethanol. [1]

$$C-C$$

  12.2 Ethanol is also produced by fermentation of sugars.
Give the conditions needed for this reaction to happen. [2]
  12.3 Describe what is seen when sodium is added to excess ethanol. [2]

13 The formula of propanoic acid is CH_3CH_2COOH.
  13.1 Explain how this shows that propanoic acid is a carboxylic acid. [2]
  13.2 Predict what is seen when excess calcium carbonate is added to propanoic acid. [1]
  13.3 Name the type of organic compound formed when propanoic acid reacts with ethanol. [1]
  13.4 **Higher Tier only:** Explain why propanoic acid is described as a **weak acid**. [2]

14 The diagram shows a structural formula for phenylethene.
  14.1 Complete the diagram to show the formation of poly(phenylethene). [3]

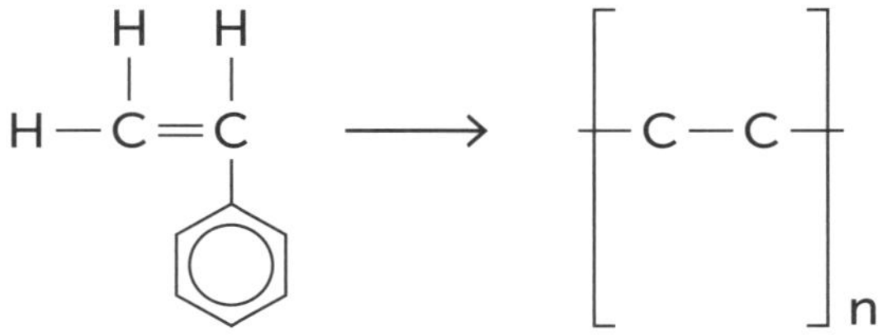

  14.2 Give a reason why poly(phenylethene) is an example of an **addition** polymer. [1]

15 DNA and starch are natural polymers.
  15.1 Describe the structure of a DNA molecule. [2]
  15.2 Name the types of monomers that form DNA and starch. [2]

Higher Tier only:

16 Polyesters are polymers.
  16.1 Explain why reactions that produce polyesters are described as **condensation** reactions. [2]
  16.2 Compare the monomers needed to produce a condensation polymer with the monomers needed to produce an addition polymer. [2]

17 The structural formula for alanine is $H_2NCH(CH_3)COOH$.
  17.1 Identify the type of organic compound represented by alanine. [1]
  17.2 Draw the repeating unit for the polypeptide formed by alanine monomers. [1]

PURE SUBSTANCES

The chemical meaning of the word **pure** is different from its everyday meaning.

Elements, compounds and mixtures

An **element** consists of atoms with the same **atomic number**. A **compound** consists of two or more different elements, chemically combined in fixed proportions as a result of a chemical reaction.

A **mixture** contains two or more elements or compounds:
- the individual substances are not chemically combined
- the individual components are not in fixed proportions.

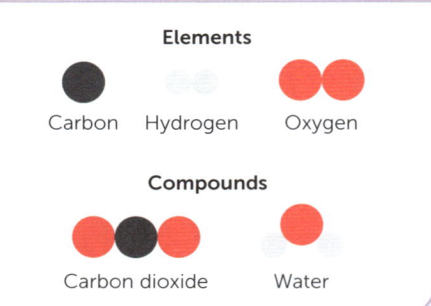

Pure vs impure substances

To a chemist, a pure substance consists of one element or one compound only, with no other element or compound mixed with it. In everyday life, a pure substance is usually a substance with nothing else added to it.

Orange juice is often described as 'pure' because the carton only contains orange juice. The juice itself is impure because it is a mixture of water, sugar, acids, 'bits' and many other substances.

Melting points and boiling points

A pure substance has a specific **melting point** and **boiling point**. An impure substance:
- melts and boils at different temperatures than a pure substance
- may melt or boil over a range of temperatures instead of at a single temperature.

You can see the difference when the temperature of a substance is recorded as it is heated until it melts or boils.

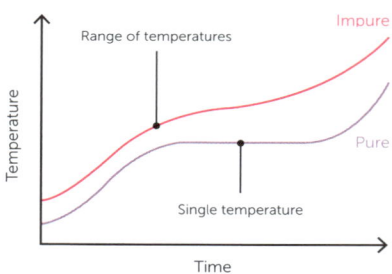

The table shows the melting points and boiling points of two samples of water.

Sample	Melting point in °C	Boiling point in °C
Distilled water	0.0	100.0
Seawater	−1.9	100.5

Explain why the seawater cannot be pure water. [2]

Distilled water is pure water.[1] *The melting and boiling points of seawater are different from those of pure water*[1]*, so it cannot be pure water.*

FORMULATIONS

A **formulation** is a mixture in which each component is there for a good reason.

Formulations vs simple mixtures

The table shows similarities and differences between formulations and simple mixtures.

	Formulation	Simple mixture
Number of components	Two or more, usually many	Two or more, often a few
Proportion of each component	Fixed	Not fixed

Examples of formulations include:

Cleaning products

Fertilisers

Foods

Fuels

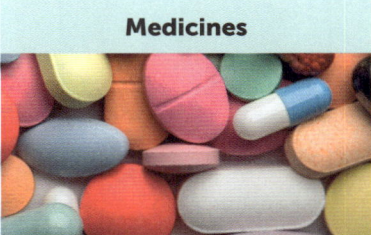
Medicines

Paints

Components of formulation

Each individual substance in a formulation has a particular purpose. They are carefully chosen and mixed in controlled quantities to produce a formulation with the desired properties.

The table shows the composition of two alloys used in jewellery.

Alloy	% Gold	% Copper	% Silver
Yellow-coloured gold	75.00	12.50	12.50
Rose-coloured gold	75.00	22.25	2.75

Explain why these alloys are formulations. [2]

> They contain different metal elements in carefully measured percentages[1] which gives them different, useful properties[1].

EXPLAINING CHROMATOGRAPHY

Chromatography separates a mixture of coloured solutes in a solution.

Phases

Chromatography relies on two 'phases':
- A **stationary phase** that does not move
- A **mobile phase** that moves past the stationary phase

Required practical activity 6 involves paper chromatography. The stationary phase is in the paper, and the mobile phase is a solvent such as water or propanone.

R_f values

A substance in a solution forms chemical bonds with both phases. The relative strengths of these bonds determine how far the substance travels up the paper with the solvent.

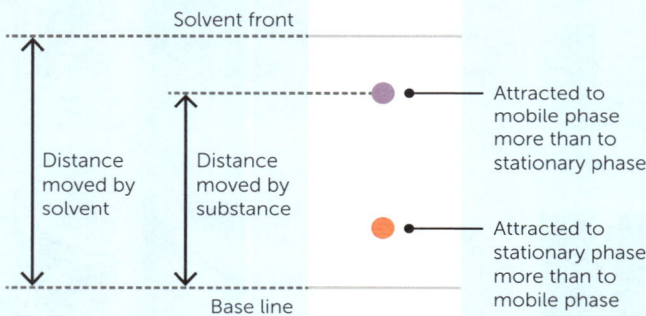

The R_f value of a substance is a measure of the relative distance it travels up the paper:

$$R_f = \frac{\text{Distance moved by substance}}{\text{Distance moved by solvent}}$$

Remember to measure from the centre of a spot to the base line.

Gas-liquid chromatography is an advanced type of chromatography in which the mobile phase is a gas, and the stationary phase is a liquid on a solid support.

---

1. Explain whether the diagram shows a chromatogram of a pure substance or a mixture. [2]
2. In the diagram, the centre of the blue spot is 72 mm from the base line. The solvent travelled 94 mm. Calculate the R_f value of the blue spot.
   Give your answer to an appropriate number of significant figures. [3]

> 1. (It is a mixture because) there are two spots[1] and a pure substance would produce one spot[1].
> 2. $R_f = \frac{72\ mm}{94\ mm}$[1] = 0.76595[1] = 0.77[1] to 2 significant figures.

REQUIRED PRACTICAL 6 (12)
Paper chromatography

This required practical activity helps you develop your ability to separate mixtures, and to accurately make and record measurements.

Separating mixtures of dyes

Food colourings are soluble in water, so water can be used as the **mobile phase** when analysing them using paper chromatography. You can revise the method and how it works on **page 7** and opposite.

Recording the results

A table is a good way to record several measurements or observations. A food colouring may contain more than one substance, so you could make a separate table for each food colouring.

The important thing is that the results are recorded clearly. In this example, the table includes a column for the R_f values after they have been calculated as part of the **analysis**.

Food colouring: Green
Distance travelled by solvent = 80 mm

Colour of spot	Distance travelled by spot in mm	R_f value of spot
Green	40	0.50
Blue	60	0.75

You can calculate the R_f value of each spot using the equation opposite.

Food colourings make food more interesting

1. Describe **two** features of identical compounds on a chromatogram. [2]
2. A student uses paper chromatography to analyse a sample of ink. Suggest **two** precautions needed when drawing the base line.
   Give a reason for each suggestion. [4]

   1. They will have the same R_f value[1] and the same colour[1].
   2. Draw the line with pencil rather than pen[1] because pencil will not dissolve in the solvent[1]. Make sure the line will be above the surface of the solvent[1] otherwise the ink will leave the paper[1].

Evaluating the investigation

In your **evaluation**, you could discuss the method and how well it worked, including any difficulties you might have had when analysing the results. You could also suggest improvements to the method, and ideas for extending the investigation.

| 4.8.2.1–4.8.2.4 | RPA7 | 5.8.2.1–5.8.2.4 |

IDENTIFYING COMMON GASES

Some gases can be identified using simple laboratory tests.

When you describe a test for a gas:
- Say what you need to do
- Say what you observe if the test is positive.

Hydrogen

Hydrogen burns rapidly in air.

To test for hydrogen:
- Hold a lighted wooden splint near the open end of a test tube
- If hydrogen is present, it burns with a 'pop' sound.

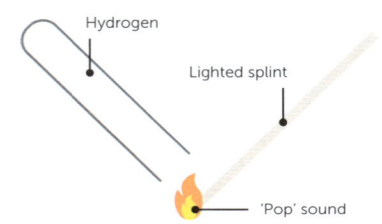

Oxygen

The greater the relative amount of oxygen present, the more rapidly a substance will burn.

To test for oxygen:
- Hold a glowing wooden splint inside a test tube
- If oxygen is present, the splint relights.

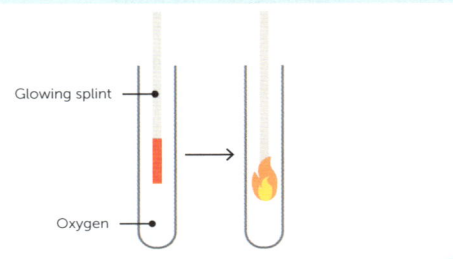

Carbon dioxide

To test for carbon dioxide:
- Shake the gas with limewater
- If carbon dioxide is present, the limewater turns milky.

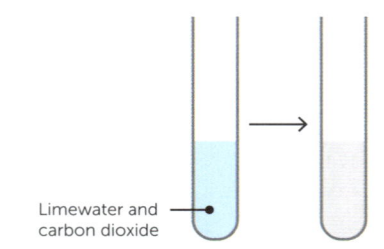

Chlorine

Chlorine has a distinctive, choking smell. It turns some coloured substances colourless.

To test for chlorine:
- Put damp litmus paper into the gas
- If chlorine is present, the litmus is bleached and the paper turns white.

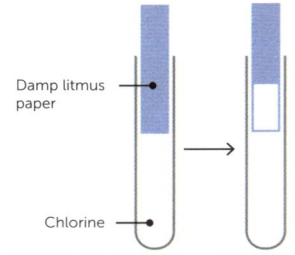

A student bubbled carbon dioxide through limewater. The limewater turned cloudy white. Explain why this reaction happened. [3]

Limewater is aqueous calcium hydroxide solution.[1] It reacted with the carbon dioxide to produce an insoluble white precipitate[1] of calcium carbonate.[1]

4.8.3.1 RPA7 Chemistry

FLAME TESTS

A **cation** is a positive ion. Some metal cations give distinctive colours to flames.

Carrying out a flame test

You can carry out a flame test on a solid sample or a sample in aqueous solution:
- Use a damp wooden splint (as in this test for calcium shown)
- Use a nichrome wire loop.

It is more difficult to use nichrome wire because the loop must be cleaned before the first test, then between each test:
1. Dip the loop in dilute hydrochloric acid.
2. Hold the loop in the hot part of a blue Bunsen burner flame.
3. Repeat steps 1 and 2 until there is little or no extra colour to the flame.

Test a sample using the cleaned nichrome wire loop:
4. Dip the loop in dilute hydrochloric acid, then in the sample.
5. Hold the loop in the edge of a blue Bunsen burner flame.
6. Observe and record the flame colour obtained.

Calcium flame test

Flame test colours

The table summarises the flame colours that you need to know. The photos show three flame tests carried out using a nichrome wire loop.

Cation	Flame colour
Li^+	Crimson
Na^+	Yellow
K^+	Lilac
Ca^{2+}	Orange-red
Cu^{2+}	Green

Lithium | Sodium | Copper(II)

A student carried out a flame test using potassium chloride.
The photo shows the results.
Suggest an explanation for why the flame colour differs from the expected colour. [3]

The flame colour should be lilac[1] but the colour is masked by another flame colour[1]. This may be because the nichrome wire was not cleaned properly first, or because the student used a mixture of ions.[1]

AQA GCSE Chemistry 8462 / 8464 – Topic 8

4.8.3.2 **Chemistry**

METAL HYDROXIDES

Some metal **cations** produce hydroxide precipitates with distinctive colours.

Precipitation reactions

Most compounds of lithium, sodium or potassium are soluble in water. Other metals may produce insoluble compounds. If these compounds form in a solution during the reaction, they appear as cloudy or jelly-like **precipitates**. For example:

- iron(II) sulfate solution reacts with sodium hydroxide solution to form a green precipitate:

$$FeSO_4(aq) + 2NaOH(aq) \rightarrow Na_2SO_4(aq) + Fe(OH)_2(s)$$

- iron(III) nitrate solution reacts with sodium hydroxide solution to form a brown precipitate:

$$Fe(NO_3)_3(aq) + 3NaOH(aq) \rightarrow 3NaNO_3(aq) + Fe(OH)_3(s)$$

Carrying out a test using sodium hydroxide solution

You must use samples in aqueous solution, so dissolve any solid samples in water first.

1. Use a clean pipette to transfer 1–2 cm³ of the test solution to a test tube.
2. Use another pipette to add a few drops of sodium hydroxide solution.
3. Observe the colour of any precipitate that forms.
4. If a white precipitate forms, add excess sodium hydroxide solution to see if it will dissolve.

Metal hydroxide colours

The table summarises the metal hydroxide precipitate colours that you need to know.

Cation	Metal hydroxide precipitate colour
Fe^{2+}	Green
Fe^{3+}	Brown
Mg^{2+}	White
Ca^{2+}	White
Cu^{2+}	Blue
Al^{3+}	White – dissolves in excess NaOH

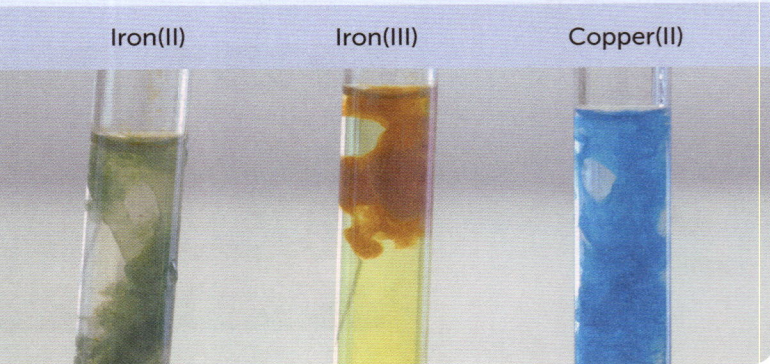

Iron(II)   Iron(III)   Copper(II)

1. Explain why potassium ions cannot be detected using sodium hydroxide solution. [2]
2. Write a balanced equation for the reaction between copper(II) chloride solution and sodium hydroxide solution. Include state symbols. [3]

   1. Potassium hydroxide is soluble in water[1] so potassium ions do not form a precipitate with sodium hydroxide solution[1].

   2. $CuCl_2(aq) + 2NaOH(aq) \rightarrow 2NaCl(aq) + Cu(OH)_2(s)$ Correct formulae[1], balancing[1], state symbols[1].

4.8.3.3–4.8.3.5 | RP7 | Chemistry

IDENTIFYING ANIONS

An **anion** is a negatively charged ion. Some anions may be identified using simple laboratory tests.

When you describe a test for an anion: say what you need to do, and state what you observe if the test is positive.

Test for Carbonate ions

Carbonate ions, CO_3^{2-}, react with acids to produce carbon dioxide.

To test for carbonates:
- Add a few drops of dilute hydrochloric acid to a solid sample or a sample in solution.
- If carbonate ions are present, bubbles of gas are given off.

It is possible that the gas is not carbon dioxide, so you need to carry out a further test to confirm that it is carbon dioxide. Limewater turns milky when carbon dioxide is bubbled through it (**page 126**).

Limewater turns milky when carbon dioxide is bubbled through it

Test for halide ions

Chloride, bromide and iodide ions form different coloured precipitates with silver ions.

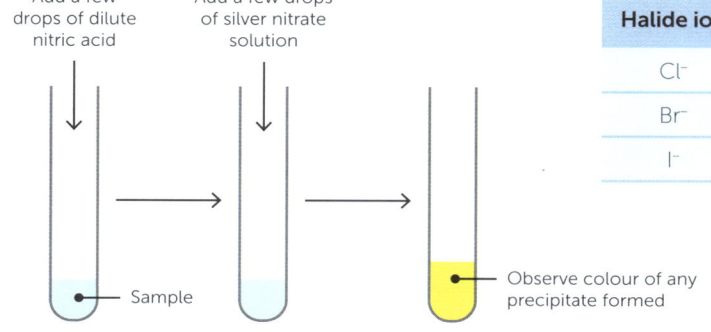

Halide ion	Colour of silver halide precipitate
Cl^-	White
Br^-	Cream
I^-	Yellow

★ To help with recalling the silver halide colours, think 'milk, cream, butter'.

1. Describe a simple laboratory test for sulfate ions, SO_4^{2-}, in an aqueous solution. [3]
2. Explain why dilute acids are added first in the halide ion and sulfate ion tests. [3]

   1. Add a few drops of dilute hydrochloric acid to the solution[1] then a few drops of barium chloride solution[1]. A white precipitate forms if sulfate ions are present.[1]
   2. Carbonate ions form white precipitates with silver ions and barium ions.[1] These reactions would give a false positive result for chloride ions or sulfate ions.[1] The acid reacts with any carbonate ions present and stops this happening.[1]

AQA GCSE **Chemistry 8462 / 8464 – Topic 8**

Chemistry

REQUIRED PRACTICAL 7
Chemical analysis

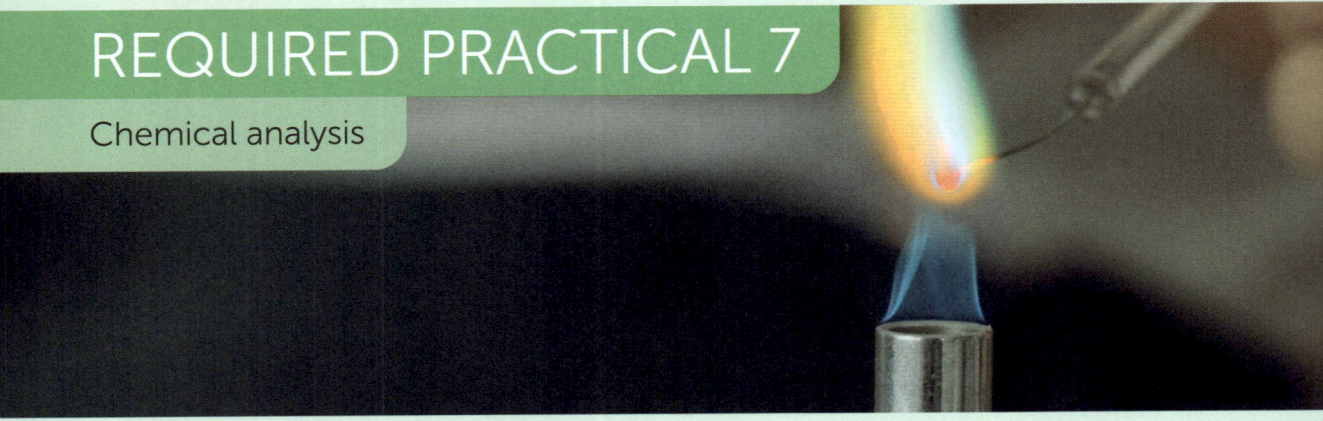

This required practical activity helps you develop your ability to use heating devices and methods safely, and to analyse and identify unknown substances.

Identifying cations

You can carry out a **flame test** on a solid sample or a sample in aqueous solution. Revise the method needed on **page 127**.

Cation	Flame colour
Li^+	Crimson
Na^+	Yellow
K^+	Lilac
Ca^{2+}	Orange-red
Cu^{2+}	Green

Li^+, Na^+ and K^+ compounds can only be identified this way.

You can identify some metal cations by the colour of their hydroxide **precipitates**. Revise the method needed on **page 128**.

Cation	Hydroxide precipitate colour
Fe^{2+}	Green
Fe^{3+}	Brown
Mg^{2+}	White
Ca^{2+}	White
Cu^{2+}	Blue
Al^{3+}	White – dissolves in excess NaOH

Identifying anions

You can identify **halides** in aqueous solution:
1. Add a few drops of dilute nitric acid.
2. Add about 1 cm³ of silver nitrate solution.
3. Record the colour of any precipitate.

Halide ion	Colour of silver halide precipitate
Cl^-	White
Br^-	Cream
I^-	Yellow

You can identify **sulfates** in aqueous solution:
1. Add a few drops of dilute hydrochloric acid.
2. Add about 1 cm³ of barium chloride solution.
3. Record whether a white precipitate forms.

You can identify **carbonates** as solid samples or in aqueous solution:
1. Add about 1 cm³ dilute hydrochloric acid.
2. If you see bubbles, use a teat pipette to transfer the gas to limewater.
3. Record whether a white precipitate forms.

---

1. Describe how to distinguish between a sodium salt, a magnesium salt and a calcium salt. [4]
2. Identify the **five** ions that give white precipitates in chemical analyses. [5]

    1. Dissolve in water and add sodium hydroxide solution.[1] The Mg^{2+} and Ca^{2+} salts give white precipitates but the Na^+ salt does not.[1] Then carry out flame tests on the Mg^{2+} and Ca^{2+} salts.[1] The Ca^{2+} salt gives an orange-red flame but the Mg^{2+} salt does not.[1]
    2. Mg^{2+}[1] Ca^{2+}[1] Al^{3+}[1] Cl^-[1] SO_4^{2-}[1]

You can revise the detailed methods for identifying halides, sulfates and carbonates on **page 129**.

Make sure you add the correct acids.

4.8.3.6 Chemistry

INSTRUMENTAL METHODS OF ANALYSIS

Instrumental methods of analysis use machines rather than test tube reactions to analyse substances.

Advantages

Compared to simple chemical tests, instrumental methods of analysis are:
- **accurate** (they correctly detect and identify substances in a sample)
- **sensitive** (they can detect very small quantities of a substance)
- **rapid** (they are very quick).

Robot arms and automatic sample injectors allow machines to analyse many samples per day, with little intervention from scientists once they have been set up. Computers are often used to analyse the results from the machines.

Flame tests are simple chemical tests. They have disadvantages, including that:
- they can only detect the presence of metal ions
- they cannot measure the **concentrations** of each ion in a sample
- their results are uncertain if a sample contains a mixture of metal ions.

Flame emission spectroscopy can distinguish between metals ions in a sample, and also measure their concentrations. You can revise this instrumental method of analysis on **page 132**.

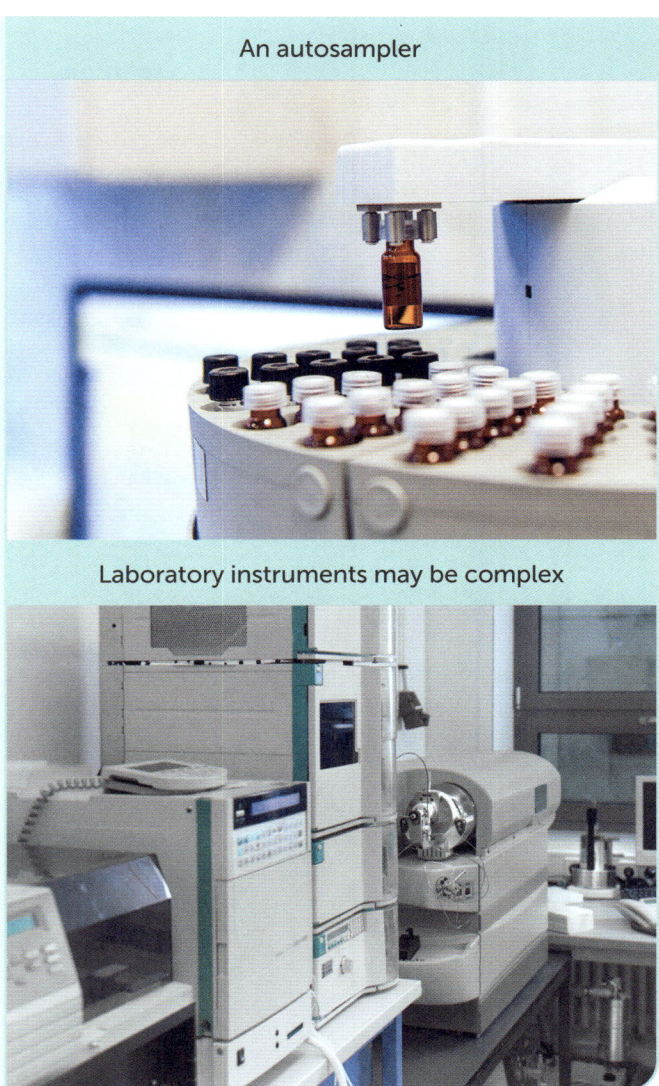

An autosampler

Laboratory instruments may be complex

Infrared spectroscopy is an instrumental method of analysis that detects the different covalent bonds in compounds. Suggest an advantage of this method compared to chemical tests for anions and cations in compounds. [2]

Flame tests and precipitation tests can only analyse ionic compounds[1]. Infrared spectroscopy can analyse molecular substances instead because they contain covalent bonds[1].

AQA GCSE **Chemistry 8462 / 8464 – Topic 8**

4.8.3.7 Chemistry

FLAME EMISSION SPECTROSCOPY

Flame emission spectroscopy is used to analyse metal ions in solutions.

An instrumental method of analysis

Flame emission spectroscopy is similar in some ways to **flame tests**:
- A sample is dissolved in water to form a **solution**
- The sample is placed in a very hot flame
- The light emitted from the flame is passed through a **spectroscope**

The spectroscope splits the colours in the light into a **line spectrum**. Different metal ions produce different lines, forming a spectrum rather like a bar code.

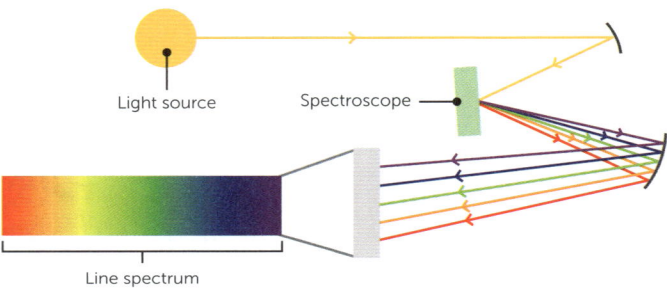

Measuring concentrations of ions

The **intensity** (brightness) of a line in the spectrum is related to the concentration of ions. A **calibration curve** shows the intensity at different, known concentrations of ions.

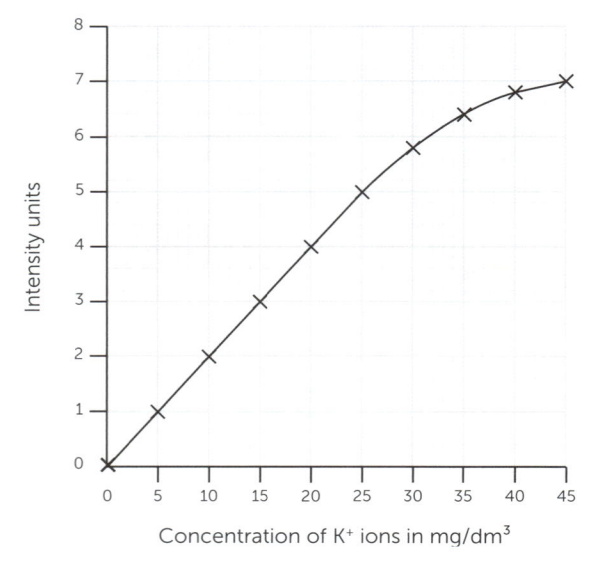

1. The diagram shows four line spectra produced by flame emission spectroscopy.

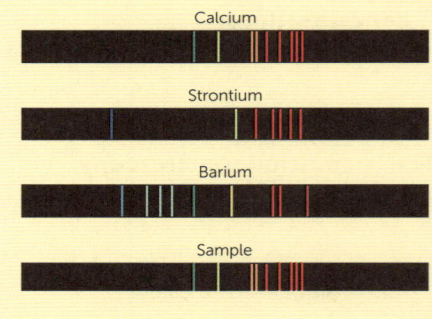

Identify the sample.
Give a reason for your answer. [2]

2. A sample containing K⁺ ions produces an intensity of 3.6 units. Determine its concentration. [1]

1. Calcium.[1] The lines produced by the sample match the lines produced by calcium.[1]

2. 18 mg/dm³[1]

TOPIC 8

EXAMINATION PRACTICE

01 Vigorous bubbling is seen when potassium iodide is added to hydrogen peroxide solution. Describe a test to confirm that the gas produced is oxygen. [2]

02 Bubbling is seen when calcium carbonate reacts with dilute hydrochloric acid. Describe a test to confirm that the bubbles contain carbon dioxide. [2]

03 A student carried out electrolysis using sodium chloride solution. Bubbles of gas were seen at each electrode. The student predicted the production of hydrogen at the cathode and chlorine at the anode. Describe how the student could test these predictions. [4]

04 Pure iron is soft but alloys of iron with carbon and other elements are much harder.
    04.1 Describe how the chemical definition of 'pure' differs from its everyday definition. [2]
    04.2 Give a reason why alloys are formulations rather than simple mixtures. [1]

05 The diagram shows the results of a paper chromatography experiment.

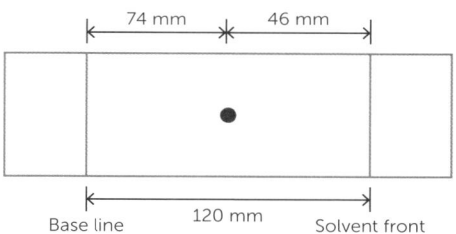

Calculate the R_f value for the substance on the chromatogram. Use the equation:

$$R_f = \frac{\text{Distance moved by substance}}{\text{Distance moved by solvent}}$$

Give your answer to an appropriate number of significant figures. [3]

06 Dodecanoic acid is a white solid at room temperature. Its melting point is 43.8 °C. Explain how a student could determine whether a sample of dodecanoic acid is pure. [2]

07 Different fertilisers are described by three numbers. For example, 5-3-3 fertilisers contain 5% nitrogen compounds, 3% phosphorus compounds, 3% potassium compounds, with the remaining ingredient being carefully chosen to give the fertiliser its desired properties.
Explain how this information tells you that fertilisers are formulations. [3]

08 Explain how paper chromatography separates mixtures. [4]

09 A student uses paper chromatography to separate the coloured substances in ballpoint pen ink. These substances are soluble in propanone but not in water.
Describe how the student could use paper chromatography to determine whether black ink contains more than one coloured substance. In your answer, include essential apparatus and methods, and how the student could use their results. [6]

Chemistry only:

10   A compound **X** consisted of two different ions. The table shows the results of two tests that were carried out on **X**.

Test	Result
Flame test	Lilac flame seen
Dilute hydrochloric acid added	Bubbling seen

Limewater turned cloudy white when the bubbles passed through it.

10.1   Name the gas that reacted with the limewater. [1]

10.2   Determine the name of compound **X**. [2]

11   A student carried out a series of tests on a solution **Y** to determine which ions are present. The table shows what the student did and the results obtained.

Test	Result
1. Flame test	Orange-red flame seen
2. Dilute nitric acid added, followed by silver nitrate solution	Yellow precipitate seen
3. Sodium hydroxide solution added to a solution of **X**	Brown precipitate seen
4. Dilute hydrochloric acid added, followed by barium chloride solution	No visible change

11.1   Identify the ions responsible for the results of Tests 1, 2 and 3. [3]

11.2   Explain what the results of Test 4 show. [2]

12   A student carried out a test to confirm the presence of chloride ions in a solution. The student added a few drops of hydrochloric acid, then a few drops of silver nitrate solution.

12.1   Explain why the student should add dilute acid in this test. [2]

12.2   Explain why the student should **not** have added dilute hydrochloric acid. [2]

13   Give **two** general advantages of instrumental methods of analysis compared to chemical tests. [2]

14   The diagram shows four line spectra produced by flame emission spectroscopy. [2]

14.1   Identify the ions present in the sample **Z**. [2]

14.2   Describe **one** advantage of flame emission spectroscopy compared to flame tests. [2]

`4.9.1.1–2` `5.9.1.1–2`

THE ATMOSPHERE

The Earth's **atmosphere** has changed over billions of years but has remained stable over the last 200 million years.

- About 4/5 nitrogen, N_2
- About 1/5 oxygen, O_2
- Smaller proportions of other gases

The early atmosphere

The Earth formed just under 4.6 billion years ago. Volcanic activity released gases such as water vapour and carbon dioxide. As the Earth cooled water vapour **condensed**, fell as rain, and formed oceans. Carbon dioxide then dissolved in the oceans.

Nitrogen was also released from volcanoes. This is a relatively **unreactive** gas, so it gradually built up in the atmosphere. Small proportions of ammonia and methane may also have been in the atmosphere.

A volcanic eruption

Comparisons

The modern atmospheres of Venus and Mars are mainly carbon dioxide, with little or no oxygen. The Earth's early atmosphere may have been like these atmospheres.

The atmosphere of Venus has strongest greenhouse effect in the Solar System. Surface temperatures reach over 460 °C. You can revise greenhouse gases and the greenhouse effect on **pages 138** and **139**.

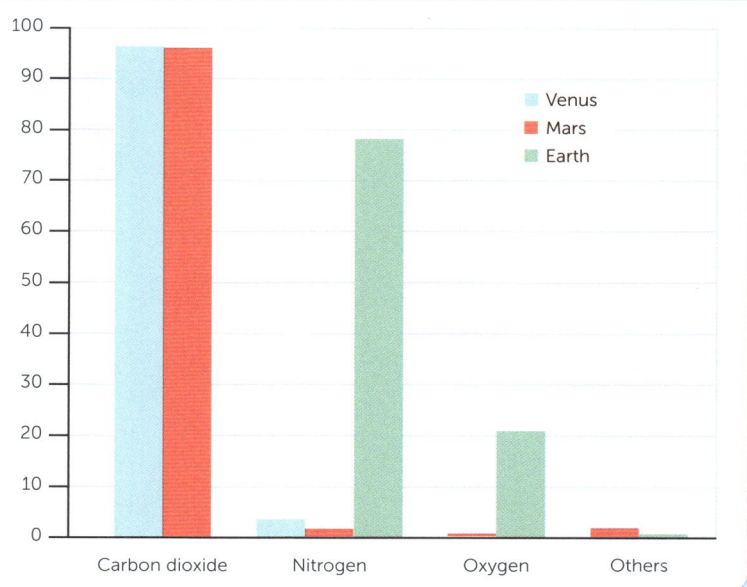

---

1. Suggest a reason why evidence for the composition of the early atmosphere is limited. [1]
2. Compare the Earth's atmosphere today with the Earth's early atmosphere. [3]

   1. It was billions of years of ago and no-one was there to observe the atmosphere then.[1]
   2. Today there is more nitrogen[1], more oxygen[1], less carbon dioxide[1].

AQA GCSE **Chemistry 8462 / 8464 – Topic 9**

HOW OXYGEN INCREASED

Photosynthesis

Photosynthesis is an important process in the development of the atmosphere. Plants and **algae** make their own food by photosynthesis. Overall:

$$\text{carbon dioxide} + \text{water} \xrightarrow{\text{light}} \text{glucose} + \text{oxygen}$$

$$6CO_2 + 6H_2O \xrightarrow{\text{light}} C_6H_{12}O_6 + 6O_2$$

These living things produce the oxygen that is in the modern atmosphere.

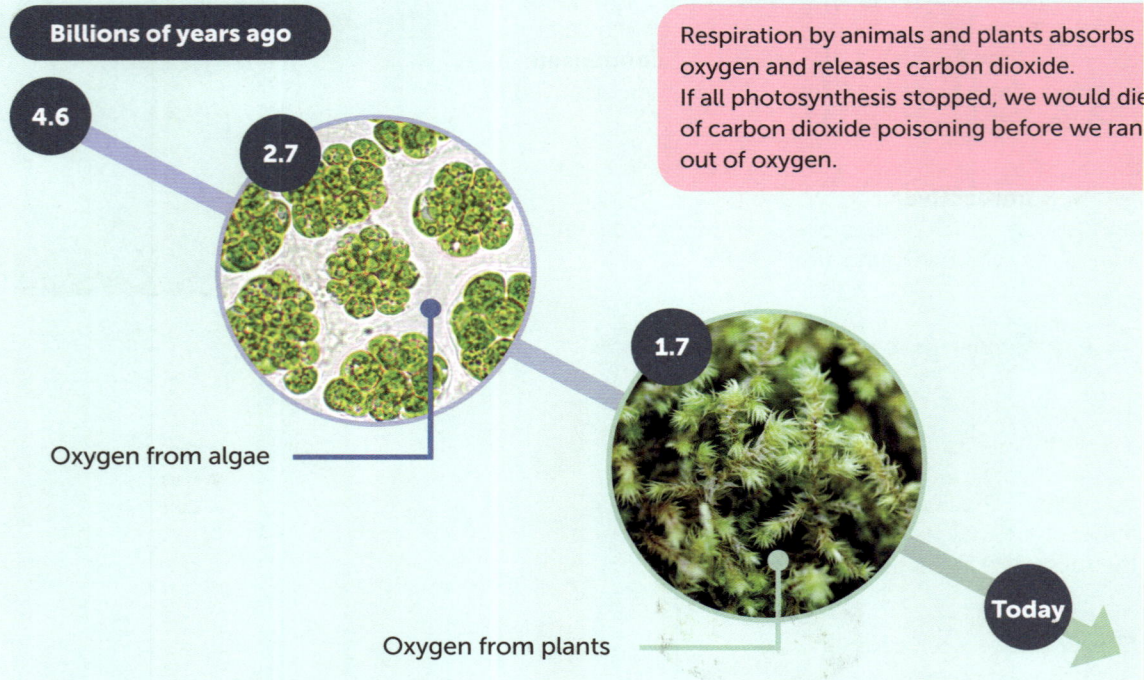

Respiration by animals and plants absorbs oxygen and releases carbon dioxide. If all photosynthesis stopped, we would die of carbon dioxide poisoning before we ran out of oxygen.

Animals could start to evolve when there was enough oxygen in the atmosphere, around 600 million years ago.

The photo shows a type of rock called banded iron. It is about 2.7 billion years old. The red layers consist of iron oxide.

(a) Suggest how banded iron provides evidence for the existence of oxygen in the atmosphere billions of years ago. [2]

(b) Describe how the oxygen was produced. [2]

(a) Iron reacted with oxygen to form the red layers.[1] The rock is about 2.7 billion years old, so oxygen must have been present billions of years ago.[1]

(b) Oxygen was produced by photosynthesis[1] in algae and plants[1].

HOW CARBON DIOXIDE DECREASED

Several processes caused the proportion of carbon dioxide to decrease.

Fossil fuels

Fossil fuels formed over millions of years from the ancient remains of living things:
- Coal formed from trees and other plants
- Crude oil and natural gas formed from plankton and algae

You can revise crude oil and natural gas on **page 104**.

The plants and algae contained carbon compounds formed from the glucose produced by **photosynthesis**. The plankton contained carbon compounds because they ate algae. This means that fossil fuels contain carbon that was absorbed from the atmosphere millions of years ago.

Coal

> There are four different types of coal, depending on how long it has been forming.

Sedimentary rocks

Carbon dioxide is a **soluble** gas. It dissolved in the Earth's early oceans, reducing its percentage in the atmosphere. Dissolved carbon dioxide releases carbonate ions, which can react with dissolved metal ions to form **insoluble** carbonates.

Over millions of years, these became **sedimentary rocks** such as limestone. You can sometimes see fossils of sea creatures that became trapped in the sediments as the limestone formed.

Fossils in limestone

Insoluble carbonates sink → Layers build up as sediments → Water squeezed out from the sediments → Sediment particles get stuck together → Sedimentary rocks form

---

1. Explain why the appearance of plants and algae decreased the percentage of carbon dioxide in the atmosphere. [2]
2. Explain how deposits of coal formed. [3]
3. **Higher Tier only:** Write an ionic equation for the production of calcium carbonate. [2]

   1. They carried out photosynthesis[1] which needs carbon dioxide absorbed from the air[1].
   2. Trees died and were buried under sediment.[1] Their remains were exposed to high pressures and temperatures[1] and were converted to coal over millions of years[1].
   3. $Ca^{2+}(aq) + CO_3^{2-}(aq) \rightarrow CaCO_3(s)$ Correct formulae and balancing[1], state symbols[1].

AQA GCSE **Chemistry** 8462 / 8464 – Topic 9

GREENHOUSE GASES

Greenhouse gases include carbon dioxide, methane and water vapour.

The greenhouse effect

Greenhouse gases keep the Earth warmer than it would be without them. They keep its temperature high enough for life to exist.

1 Energy is transmitted from the Sun to the Earth by **radiation**.

2 Some radiation is re-radiated into space by the Earth's atmosphere, clouds, and surface.

3 The remaining radiation is absorbed by the Earth's surface and warms it up.

4 The warm surface emits radiation, which warms the atmosphere.

5 Greenhouse gases in the atmosphere absorb radiation.

6 Absorbed radiation is emitted in all directions, keeping the atmosphere and the Earth's surface warm.

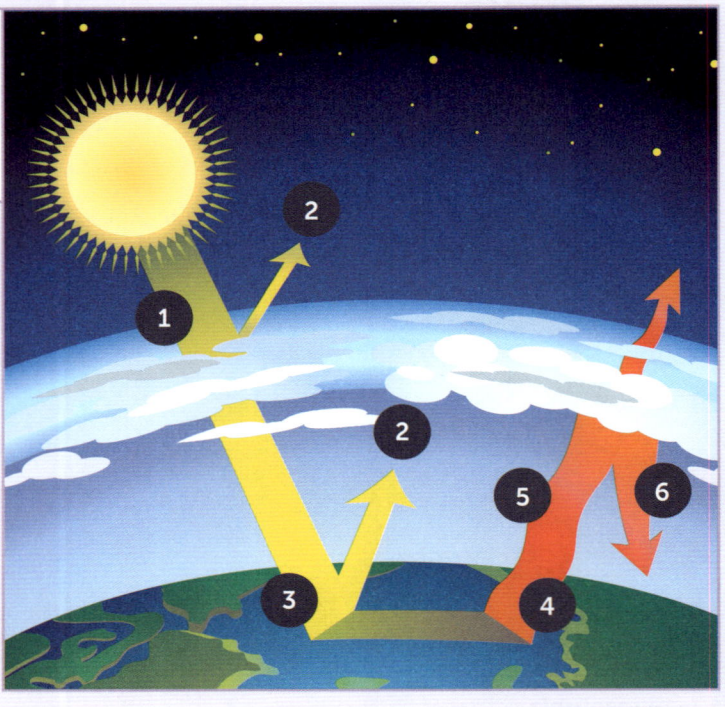

Nitrogen and oxygen comprise about 99% of the atmosphere, but they are not greenhouse gases.

 Take care not to mention the ozone layer when discussing greenhouse gases and the greenhouse effect.

In terms of wavelength, describe the radiation that passes through the atmosphere, and the radiation emitted by the Earth's surface and atmosphere. [2]

The radiation that passes through the atmosphere is short wavelength radiation.[1] *The radiation that is emitted by the Earth's surface and atmosphere is long wavelength radiation.*[1]

4.9.2.2 5.9.2.2

HUMAN ACTIVITIES AND GREENHOUSE GASES

Human activities release additional quantities of greenhouse gases into the atmosphere.

Carbon dioxide

Carbon dioxide accounts for most of the greenhouse gas emissions because of human activities, such as:
- burning fossil fuels, for example for transport and generating electricity
- cement manufacture.

Methane

Methane is a much more powerful greenhouse gas than carbon dioxide. Human activities that release methane include:
- cattle farming
- coal mining, and oil and gas production.

Methane is also released from decaying animal and plant waste on farms and in landfill sites.

Linking greenhouse emissions to global warming

As carbon dioxide levels in the atmosphere have increased, the average global temperature has increased.

The Intergovernmental Panel on Climate Change (IPCC) makes scientific conclusions based on large amounts of scientific research. It believes that it is highly likely that human activities are the main cause of global warming.

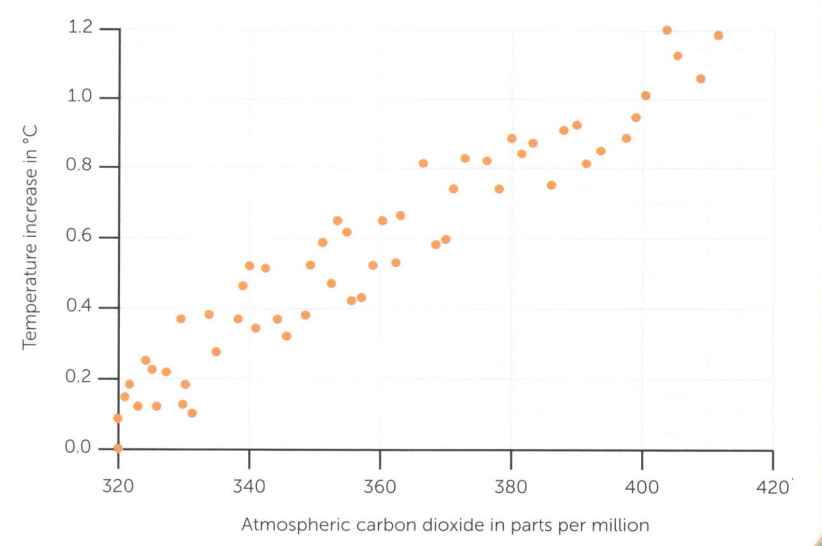

1. Describe the relationship shown in the graph. [2]
2. The IPCC uses 'peer reviewed' scientific evidence. Describe what this means. [2]

   1. There is a positive correlation[1] between the increase in temperature and the atmospheric carbon dioxide concentration[1].
   2. Peer reviewed evidence is evaluated by other scientists.[1] This means that findings reported by a scientist or research group are checked before being accepted or rejected.[1]

AQA GCSE **Chemistry 8462 / 8464 – Topic 9**

GLOBAL CLIMATE CHANGE

Increasing global temperatures are a major cause of **climate change**.

Impacts of climate change

The **weather** describes atmospheric conditions over a short period, such as a few hours or a day. Climate is weather in an area over a long period, such as 30 years. Many scientists believe that **global warming** will lead to **global climate change** – changes in the long-term weather conditions across the whole world.

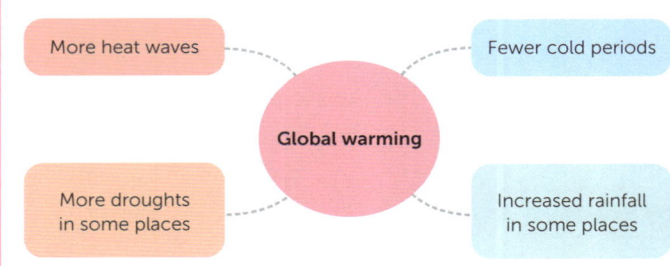

The Thames Barrier protects London from flooding during very high tides

---

1. Explain **two** problems that climate change may cause farmers. [3]
2. Climate change is leading to increased sea levels.
   (a) Give **two** reasons why this is happening. [2]
   (b) Describe **one** problem caused by increased sea levels. [2]

1. Changes in temperature and rainfall[1] may mean that some crops can no longer grow[1]. Farmers may need to plant different crops to cope with these changes.[1]
2. (a) As the oceans warm up, they expand.[1] Water from melting ice in the Antarctic and in glaciers flows into the oceans.[1]
   (b) Land near the coast will become flooded.[1] This will reduce the amount of land available for dwellings and farms.[1]

Climate change is predicted to affect different parts of the world in different ways.

Uncertainties

It is difficult for scientists to develop accurate climate **models**.
- Changes in conditions in the atmosphere are very complex.
- Scientists cannot collect evidence from every place in the world.
- The effects of global warming may be more severe in some parts of the world.

Climate models become more accurate as scientists include more factors and data.

3. Suggest **two** reasons why some people do not believe that climate change is happening. [2]

3. They may only think about parts of the evidence[1], which may be biased[1].

CARBON FOOTPRINT

A person's **carbon footprint** is a measure of their greenhouse gas emissions.

Carbon dioxide equivalent

Carbon dioxide and methane are **greenhouse gases** that contain carbon. They are released from human activities such as the production and combustion of **hydrocarbon** fuels. Sulfur dioxide and oxides of nitrogen do not contain carbon, but they are greenhouse gases. They are also released because of hydrocarbon fuels. You can revise these **pollutants** on **page 142**.

When a carbon footprint is calculated, it takes into account the greenhouse effect of these gases relative to carbon dioxide. It gives the total amount of carbon dioxide and other greenhouse gases given off over the lifetime of products, services and events. The chart shows examples.

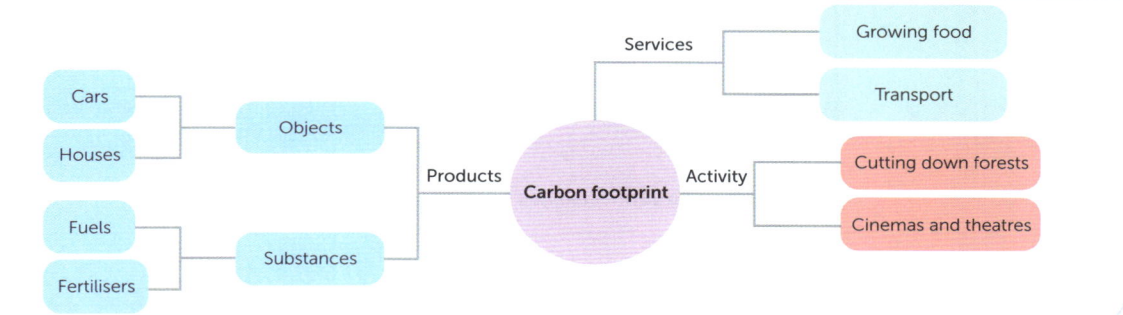

Reducing the carbon footprint

A carbon footprint is reduced if emissions of carbon dioxide and methane are reduced. This can be done by individuals such as you, by manufacturers, and by the actions of governments.

For example, you could:
- walk or cycle instead of taking journeys in cars
- make changes to your diet, such as reducing the quantity of meat you eat while maintaining a healthy diet.

Manufacturers can adopt new methods and products that use less energy and raw materials. **Life cycle assessments** help them to make decisions. (See **page 150**.)

1. Suggest **two** ways in which governments can encourage reductions in carbon footprint. [2]
2. Suggest **two** reasons why the actions of individuals to reduce their carbon footprint may be limited. Give suitable examples in your answer. [4]

   1. They can pass laws to ban or control the use of products, services or events that release a lot of greenhouse gases.[1] They can reduce taxes on things with low carbon footprints.[1]
   2. People may be reluctant to make changes[1] or they may forget to make the changes needed to reduce their carbon footprint[1]. For example, they may still want to fly to foreign countries for their holidays[1] or may forget to turn off the lights when they leave a room[1].

AQA GCSE Chemistry 8462 / 8464 – Topic 9

ATMOSPHERIC POLLUTANTS FROM FUELS

Combustion of fuels

Atmospheric **pollutants** are harmful substances released into the air. The **combustion** of **hydrocarbon** fuels releases several substances.

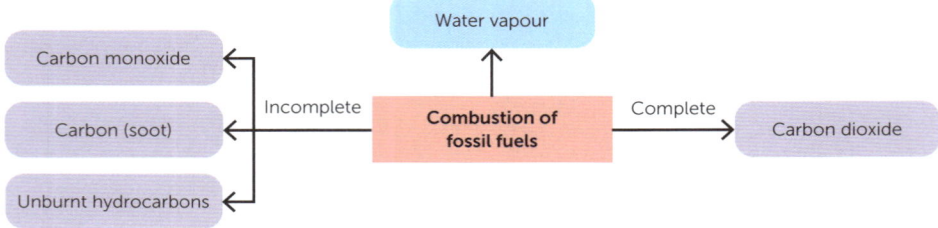

These substances can act as pollutants.

Pollutant	Properties	Problem
Carbon dioxide	**Greenhouse gas**	Increased levels lead to **global warming**
Carbon monoxide	Odourless, colourless, **toxic**	Causes breathing difficulties and even death
Carbon / unburnt hydrocarbons	Form soot and other **particulates**	Causes breathing problems and **global dimming**

Some fuels contain some sulfur. This reacts with oxygen when the fuel is used, forming sulfur dioxide. This gas causes breathing difficulties. It also dissolves in clouds to form an acidic solution that falls as **acid rain**. This harms trees and crops, and living things in rivers and lakes.

1. Oxides of nitrogen, NO_x, are a cause of acid rain.
   (a) Give **one** other problem caused by oxides of nitrogen. [1]
   (b) Explain how oxides of nitrogen form when hydrocarbon fuels are used. [2]
2. Suggest why carbon monoxide is difficult to detect. [1]

1. (a) Breathing difficulties in people.[1]
   (b) Oxygen and nitrogen in the air[1] react together at high temperatures in engines[1].
2. You cannot smell carbon monoxide or see it.[1]

TOPIC 9

EXAMINATION PRACTICE

01 In a petrol engine, a mixture of petrol and air is ignited under pressure.

Some experimental car engines use hydrogen instead of petrol. They work in the same way as petrol engines.

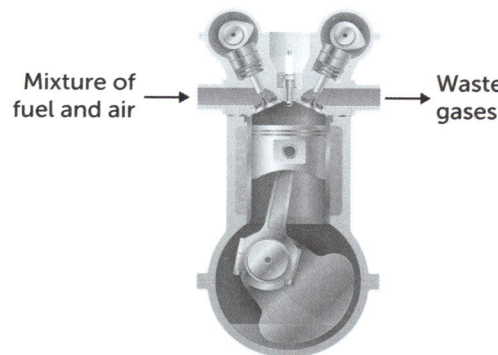

Which **two** waste gases will be produced by these hydrogen engines? Tick **two** boxes. [2]

Carbon dioxide ☐
Carbon monoxide ☐
Oxides of nitrogen ☐
Sulfur dioxide ☐
Unburned hydrocarbons ☐
Water vapour ☐

02 The pie chart shows the proportions of the two main gases in the atmosphere of Venus today.

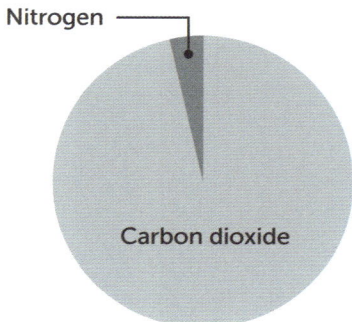

02.1 Give **two** ways in which the atmosphere of Earth and Venus today are similar. [2]

02.2 Give **one** way in which the atmosphere of Earth and Venus today are different. [1]

03 Give the approximate proportions of nitrogen and oxygen in today's atmosphere. [2]

04 One theory about the formation of the Earth's early atmosphere involved volcanoes. Describe how volcanic activity led to oceans forming. [2]

05 Write a word equation for photosynthesis. [2]

06 Describe how the presence of plants and algae changed the proportions of carbon dioxide and oxygen in the Earth's early atmosphere. [3]

07 Carbon dioxide is a greenhouse gas.
07.1 Name **one** other greenhouse gas. [1]
07.2 Give **one** human activity that increases the amount of carbon dioxide in the Earth's atmosphere on a global scale. [1]
07.3 Describe how greenhouse gases keep temperatures high enough for life on Earth. [3]

08 Describe briefly **two** possible effects of global climate change. [2]

09 Some people are very concerned about 'carbon footprint'.
09.1 Describe what is meant by a 'carbon footprint'. [2]
09.2 Suggest **one** way in which an individual person may reduce their carbon footprint. [1]

10 Hydrogen and methane, CH_4, can replace petrol in some vehicle engines. Name **one** product of the combustion of methane that is not a product of combustion of hydrogen. [1]

11 Propane, C_3H_8, is a hydrocarbon fuel used for heating and cooking. Predict **three** products of incomplete combustion of propane. [3]

12 Give **two** reasons why carbon monoxide is not easily detected. [2]

13 Name the type of pollutant that is a cause of global dimming. [1]

14 Pollutants caused by the use of hydrocarbon fuel include sulfur dioxide and oxides of nitrogen. These pollutants can cause health problems for humans, and can cause damage to the environment.

Describe how sulfur dioxide and oxides of nitrogen are formed, and the harm they cause to our health and the environment. [6]

SUSTAINABLE DEVELOPMENT

Sustainable development means meeting our needs today without making it difficult or impossible for people in the future to meet their needs.

The Earth's resources

The Earth's crust, oceans and atmosphere provide us with the **resources** we need for food, shelter, warmth and transport. Farming can provide us with more of some of these resources. Natural resources can be processed to provide energy and new materials.

Farming plants for biodiesel and bioethanol fuels competes with farming for food.

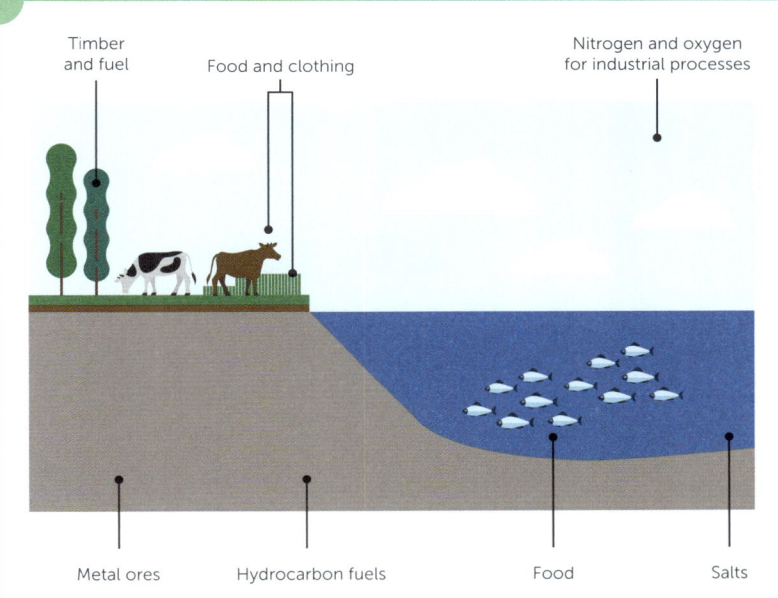

Finite and renewable resources

Metal ores, and fossil fuels such as crude oil and natural gas, are **finite resources.** They are no longer being made or are being made very slowly. This means that they will run out one day if we keep on using them.

Renewable resources are being made quickly enough to be replaced as we use them. They include biodiesel, trees and other types of **biomass**, and energy resources such as wind power and solar power.

1. Natural materials can be replaced by artificial materials made by chemical processes. Give **one** example of this type of replacement for:
   (a) material for clothing [2]
   (b) construction materials for buildings. [2]
2. Give **one** example of a natural food that can be replaced by manufactured food. [2]

   1. (a) Cotton and wool[1] can be replaced by artificial polymers such as nylon[1].
      (b) Wood from trees[1] can be replaced by bricks and steel[1].
   2. Meat from animals[1] can be replaced by mycoprotein from fungi grown in fermenters[1].

AQA GCSE **Chemistry** 8462 / 8464 — Topic 10

POTABLE WATER

Potable *[POE-tuh-bull]* water is water that is safe to drink. Unlike **pure** water, potable water contains dissolved substances, including mineral **ions** and gases.

Treating fresh water

Most potable water in the UK comes from ground water, and water in rivers and lakes. This fresh water comes from rainfall. It must be treated to make it potable.

Chlorine or ozone is bubbled through the water to sterilise it. Ultraviolet light may also be used.

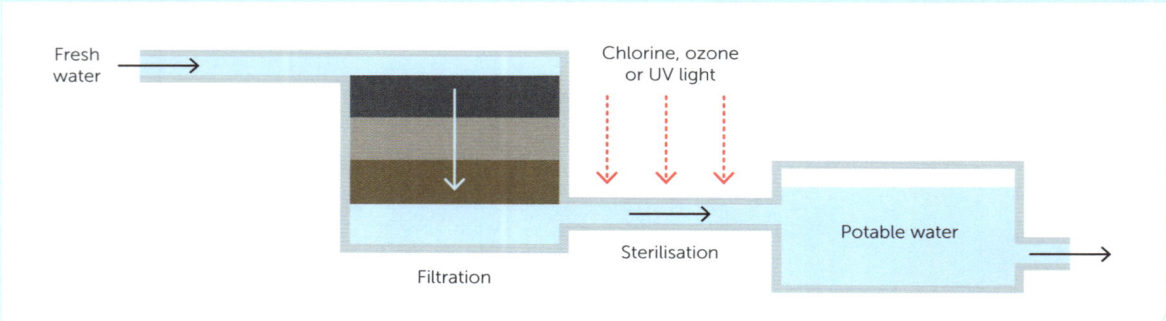

Treating salty water

Salty water and sea water contain dissolved salts that make them unsafe to drink. These **salts** are removed by desalination.

Desalination may be used if ground water supplies are limited. It can be carried out using:
- simple **distillation**
- or **reverse osmosis**.

In biology, osmosis involves water moving through a **partially permeable membrane**. In reverse osmosis, water is forced through a partially permeable membrane. The membrane lets water molecules through but not larger molecules or ions, bacteria, viruses or insoluble particles.

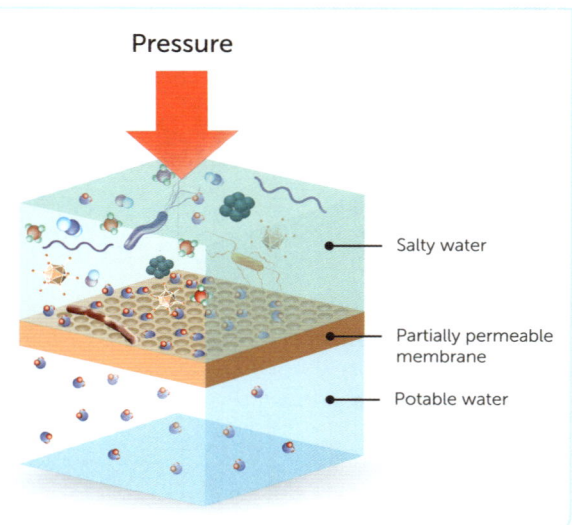

---

1. Fresh water is treated to make it potable by filtration and sterilisation. Give reasons for these **two** processes. [2]
2. Suggest **one** reason that explains why desalination is usually only carried out if fresh water supplies are limited. [1]

   1. Filtration removes insoluble particles[1] and sterilisation kills harmful microbes[1].
   2. Distillation and reverse osmosis need a lot of energy[1].

Chemistry

REQUIRED PRACTICAL 8

Analysing and purifying water samples

This required practical activity helps you develop your ability to measure pH, to heat substances safely, and to separate and purify them.

Analysing water samples

You will have some samples of impure water to test. Estimate their **pH** using **universal indicator paper** and a pH colour chart. You can then determine the mass of dissolved solids in each sample:

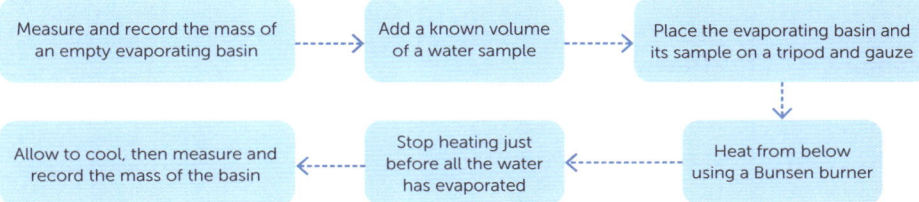

Purifying a water sample

You should use **simple distillation** to obtain pure water from one of the samples. This diagram shows one way to carry out this separation method. **Page 8** shows a more complex way to do it.

You can check the purity of your **distillate** by measuring its **boiling point**. Heat a sample of a distillate in a boiling tube until it boils. Pure water boils at 100 °C.

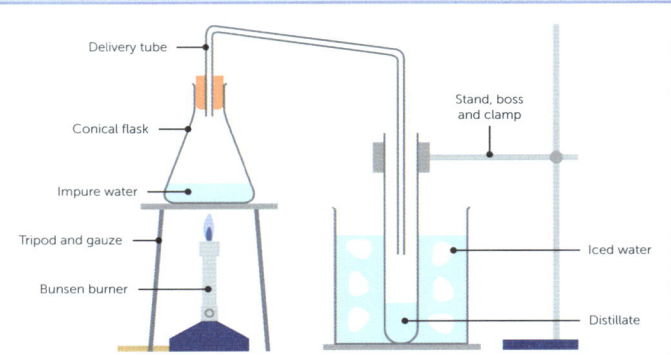

1. Suggest why a sample of water may **not** have a pH of 7. [2]
2. Describe how to use the results to calculate the mass of dissolved solids in a sample. [1]
3. Suggest **one** reason why a thermometer may not show 100 °C in boiling distilled water. [1]

   1. It contains dissolved substances[1] which dissolve to form an acidic or alkaline solution[1].
   2. Calculate: (mass of basin and contents at end) − (mass of empty basin at start).[1]
   3. The thermometer may not be calibrated properly.[1]

AQA GCSE **Chemistry** 8462 / 8464 − Topic 10

4.10.1.3 5.10.1.3

WASTE WATER TREATMENT

Waste water must be treated before it can be released back into rivers.

Waste water

Large volumes of waste water come from sources such as farms, homes and factories.

Depending on the source, this water may contain harmful substances, solids containing **organic compounds**, and harmful **microbes**. These will damage the environment and may cause disease unless the waste water is treated to remove them.

Treatment

Waste water goes through several stages before it is released into the environment.

Solid materials such as toilet paper and grit are removed. Sewage **sludge** and liquid **effluent** are produced by sedimentation, then digested by bacteria. Further biological treatment removes excess nitrates and phosphates that can cause excessive, harmful growth of algae in rivers.

1. Describe what is meant by:
   (a) anaerobic digestion   [2]
   (b) aerobic biological treatment.   [2]
2. Suggest why it is easier to obtain potable water from fresh water rather than from waste water.   [1]

1. (a) Breaking down substances[1] in the absence of air or oxygen[1].
   (b) Treatment that uses living things such as bacteria[1] in the presence of air or oxygen[1].
2. Many more stages are needed to treat waste water to make it safe to drink.[1]

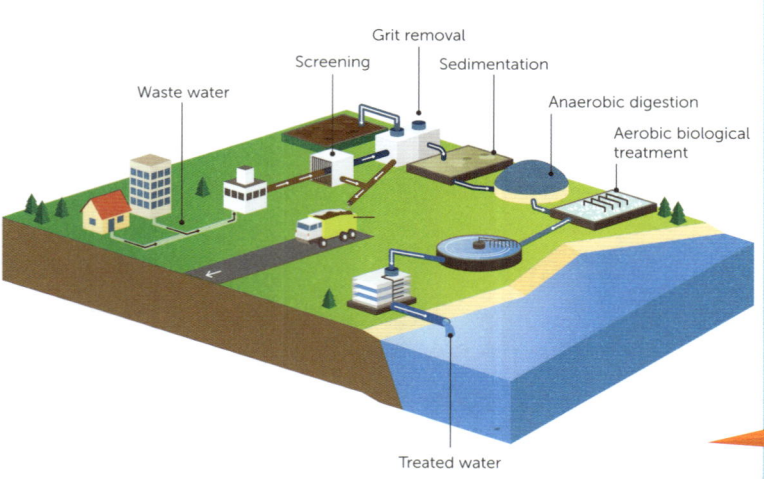

Methane produced by anaerobic digestion can be used as a fuel. You can revise hydrocarbon fuels on **pages 104** and **105**.

ALTERNATIVE METHODS OF EXTRACTING METALS

Alternative extraction methods avoid mining, transporting and disposing of vast amounts of rock.

Phytomining

Phytomining involves plants. It is useful for obtaining copper from low-grade ores.

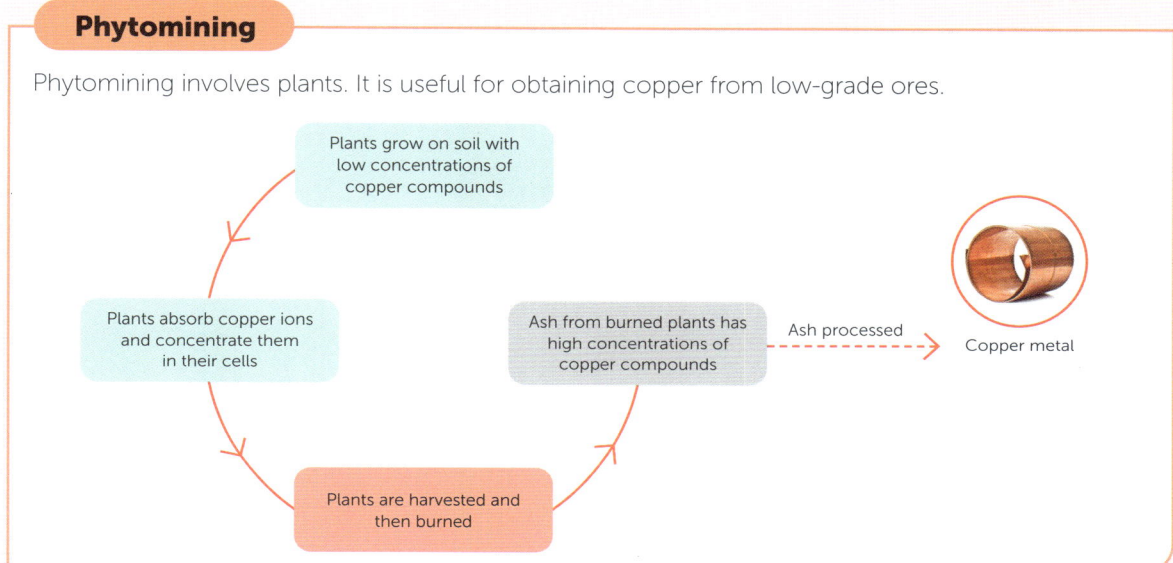

Bioleaching

Bioleaching involves bacteria. Some bacteria can absorb copper compounds and obtain energy from them through chemical reactions. These bacteria produce a solution called a **leachate**. It has copper compounds in high enough concentrations to make processing worthwhile.

Processing copper solutions

Solutions of copper compounds can be processed to extract copper in two ways:
- **Electrolysis** (See electrolysis on **pages 73–77**).
- **Displacement** reactions using scrap iron (See the reactivity series on **page 60**).

Bioleaching avoids environmental damage due to mining, but it produces acidic wastes that can also harm the environment.

---

1. Explain why low-grade ores, rather than high-grade ores, are being increasingly used. [3]
2. Suggest **one** disadvantage of using alternative methods of extraction. [1]
3. **Higher Tier only:** Write an ionic equation for the displacement of copper by iron. [3]

  1. High-grade ores contain relatively high concentrations of metals[1] but are limited resources and are becoming scarce[1]. Low-grade ores with low concentrations are what is left.[1]
  2. They are slower compared to traditional methods such as mining ores.[1]
  3. $Cu^{2+}(aq) + Fe(s) \rightarrow Cu(s) + Fe^{2+}(aq)$ Correct formulae[1], balanced[1], state symbols[1].

LIFE CYCLE ASSESSMENT

A **life cycle assessment** (**LCA**) assesses the **environmental impact** of four different stages in the lifetime of a product, and of the transport and distribution used between them.

Aims

The aims of carrying out an LCA include researching alternative methods of manufacture, maintenance, and disposal. This includes adapting the designs and sources of energy for a product.

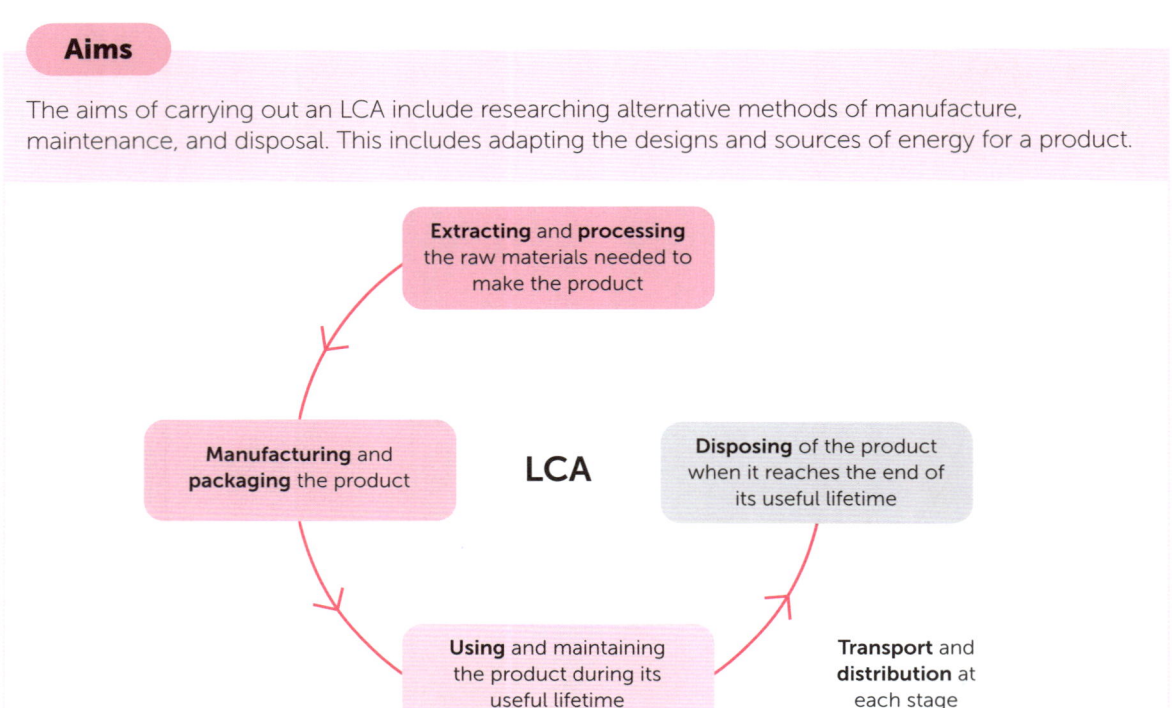

Limits of an LCA

Some aspects of an LCA can be **quantified** (given values). These can include the amount of energy used, the volume of water used, the mass of raw materials and similar resources, and some waste substances produced.

Some aspects cannot easily be quantified, such as the effects of some pollutants released because of a product. The effects of one pollutant may need to be weighed against the effects of another pollutant. Some judgement is needed, so LCAs are not entirely objective.

 You must be able to carry out and compare simple LCAs for plastic and paper shopping bags.

1. Suggest why LCAs are sometimes described as assessing the 'cradle to grave' environmental impacts of a product. [2]
2. Describe how life cycle assessments might be misused. [2]

   1. A product is assessed from when the raw materials needed to make it are extracted and processed (its 'cradle')[1] to when the product reaches the end of its life (its 'grave')[1].
   2. Manufacturers might use part of an LCA or choose a more favourable LCA[1] in order to support claims about their product or processes in advertising[1].

REUSE AND RECYCLING

Recycling can reduce the use of limited resources and reduce waste.

Raw materials and energy

The Earth's crust, oceans and atmosphere are the sources of the raw materials needed to make materials and products. Most raw materials and energy resources are **limited resources**, so it is important to conserve them. We can all help by:
- using less of these resources
- **re-using** products where possible, and
- recycling products and their parts when they have reached the end of their useful lives.

Doing this reduces harm to the environment caused by mining and quarrying. It also reduces waste.

Waste being dumped at a landfill site

Recycling

Some products can be re-used. For example, glass bottles may be returned for washing and refilling, but most products cannot be re-used. They should be recycled instead.

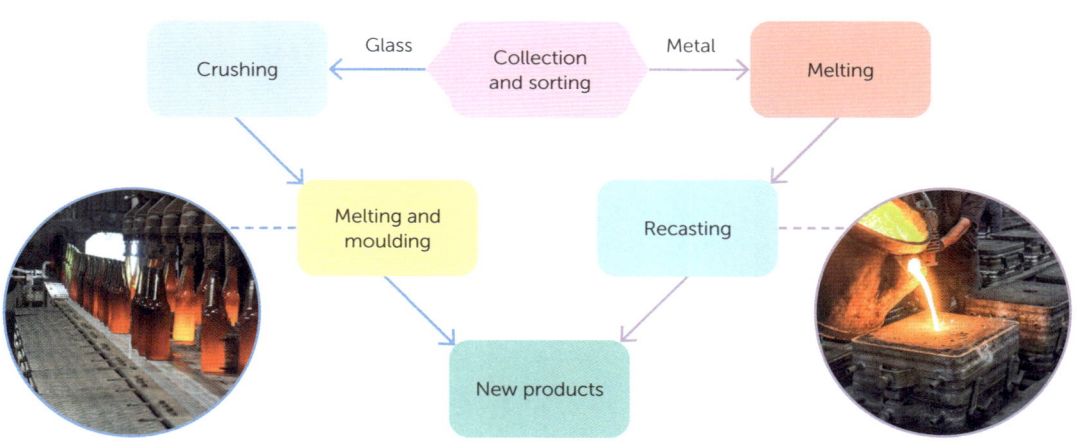

Used products usually need sorting in order to separate different materials from one another. Some contain many different materials which may be more difficult to separate.

---

1. Iron is extracted from iron ore in a blast furnace. Some scrap steel may be added to the liquid iron. Explain why this is done. [2]
2. Suggest an explanation why broken bathroom tiles are more difficult to recycle than plastic shampoo bottles. [2]

   1. It recycles used steel[1] which reduces the amount of iron ore that must be used[1].
   2. Tiles are made from clay ceramics which cannot be melted and reformed[1] but most plastics can be melted and reformed into new products[1].

4.10.3.1 Chemistry

CORROSION AND ITS PREVENTION

Chemical reactions with substances in the environment can damage materials.

Corrosion

Corrosion involves metals reacting with substances in the environment. These are usually air or water, but they can be substances such as sulfur dioxide. Corrosion starts on the surface, but eventually it can destroy the metal.

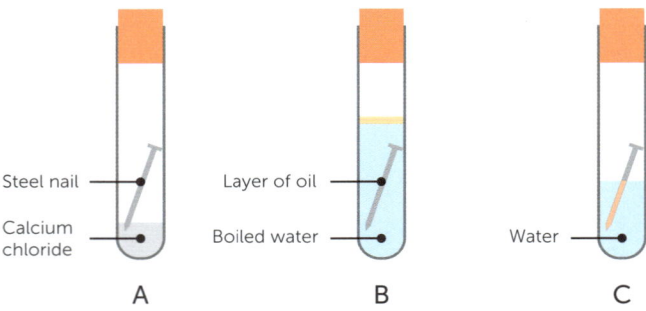

Corrosion of iron and steel

Rusting is the corrosion of iron and steel. It needs both air (oxygen) and water to happen. You can show this using a suitable experiment, such as the one below.

Tube A: Steel nail, Calcium chloride
Tube B: Steel nail, Layer of oil, Boiled water
Tube C: Steel nail, Water

The table shows the results of this experiment after a few days.

Tube	A	B	C
Air	✓	✗	✓
Water	✗	✓	✓
Nail rusts?	✗	✗	✓

1. (a) Explain why calcium chloride is added to Tube A. [2]
   (b) Explain the conditions in Tube B. [2]

1. (a) Calcium chloride absorbs water[1], so the air in Tube A is dry[1].
   (b) Air leaves water when the water is boiled[1]. The layer of oil stops air getting back into the water during the experiment[1].

Preventing corrosion

Corrosion can be prevented in different ways.

Barriers

A metal will not corrode if it has a coating that acts as a barrier. An iron or steel object will not rust if air, water, or both air and water are kept away from its surface. This is why bike chains do not rust if they are oiled regularly, and steel bike frames do not rust if they are painted.

Metal objects can be protected by **electroplating** them. This process uses **electrolysis** to apply a very thin metal coating on the surface of another metal. The inside of steel food cans can be electroplated with a thin layer of tin, which keeps air and water away from the steel.

Steel parts are electroplated with chromium to stop them rusting and to improve their appearance

Sacrificial protection

The **reactivity series** shows the relative reactivity of metals (See **page 60**).

A more reactive metal reacts with air and water more readily than a less reactive metal does. This means that a suitable metal can 'sacrifice itself' to protect iron and steel from rusting.

Magnesium and zinc are suitable **sacrificial metals**. As long as they are in contact with an iron or steel object, they corrode instead of the object. The sacrificial metal is replaced before it completely corrodes away.

Most reactive

Potassium
Sodium
Lithium
Calcium
Magnesium
Zinc
Iron
Copper

Least reactive

Sacrificial protection of a pipeline

Sacrificial protection of a ship's hull

2. Aluminium is more reactive than zinc. Explain why it does not corrode in air or water. [2]
3. A farmer's steel gate is galvanised with zinc to stop it rusting. Explain how this works. [2]

2. Aluminium has a natural, thin coating of aluminium oxide[1] which stops air and water reaching the metal below[1].

3. The zinc acts as a barrier[1] and it also acts as a sacrificial metal[1].

Food cans are commonly lined with a polymer coating to provide a further barrier against rusting.

AQA GCSE **Chemistry 8462 / 8464 — Topic 10**

4.10.3.2 Chemistry

ALLOYS AS USEFUL MATERIALS

Most metals in everyday use are mixtures of metals called **alloys**.

Copper alloys

Copper alloys include **brass** and **bronze**. They have different compositions, and their physical properties are different. This makes them suitable for different uses.

	Brass	Bronze
Composition	Copper + zinc	Copper + tin
Appearance	Gold-coloured	Reddish-brown
Typical property	**Malleable** (can be shaped easily)	Hard, resists **corrosion**
Typical uses	Plumbing parts, brass instruments	Sculptures, bells

Gold alloys

Gold is shiny and it resists corrosion, so it is often used in jewellery. Jewellery gold is usually an alloy with copper, silver and zinc.

The proportion of gold in gold alloys is measured in **carats**:
- 24 carat gold is 100% gold
- 18 carat gold is 75% gold.

Iron alloys

Iron alloys containing specific percentages of other metals and carbon are called **steels**. In general, the greater the percentage of carbon, the stronger but more brittle the metal becomes. Stainless steels (iron alloys containing chromium and nickel) are hard and resist corrosion.

You can revise structure and bonding in metals and alloys on **page 37**.

1. Explain why aluminium alloys are used in aircraft parts. [2]
2. Explain **one** use of:
   (a) Low carbon steel. [2]
   (b) High carbon steel. [2]

1. Aluminium alloys have low densities[1], so parts made from them are lightweight[1].
2. (a) Low carbon steel is used for car body panels[1] because it is easily shaped[1].
   (b) High carbon steel is used for springs and wires[1] because it is strong[1].

USING POLYMERS

Polymers have different properties depending on how they are made.

Thermosoftening and thermosetting polymers

There are two main types of polymer:
- **Thermosoftening** polymers melt when they are heated, and are more easily recycled.
- **Thermosetting** polymers do not melt when they are heated, and may blacken and char instead.

The two types of polymer have different structures and bonding.

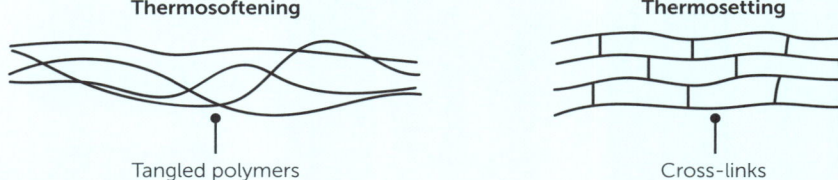

	Thermosoftening	Thermosetting
Bonding within a molecule	covalent bonding	covalent bonding
Bonding between molecules	intermolecular forces	covalent bonding
Arrangement of molecules	may be tangled	regular

Intermolecular forces are relatively weak. Little energy is needed to break them, so thermosoftening polymers melt when they are heated. Covalent bonds are strong. A lot of energy is needed to break them, so thermosetting polymers do not melt when they are heated.

Poly(ethene)

There are different types of poly(ethene):
- low-density (LD) poly(ethene)
- high-density (HD) poly(ethene).

Both are made from ethene monomers, but the reaction conditions used to make them are different. LD poly(ethene) molecules have many more branches than HD poly(ethene) molecules.

---

1. LD Poly(ethene) melts at about 115 °C but HD poly(ethene) melts at about 135 °C. Suggest why there is a difference. [2]
2. Poly(propene) is about three times stronger than poly(ethene). Give **one** reason why the two polymers have different physical properties [1]

   1. The intermolecular forces in LD poly(ethene) are weaker than those in HD poly(ethene)[1], so less heating is needed to overcome them and to separate the molecules[1].
   2. They are made from different monomers (propene and ethene).[1]

4.10.3.3 Chemistry

CLAY CERAMICS AND GLASS

The manufacture of glass and of clay ceramics involves high temperatures.

Glass

Glass is made by heating a mixture of silica sand and other substances, then cooling the liquid rapidly. The different mixtures produce different types of glass, with different properties.

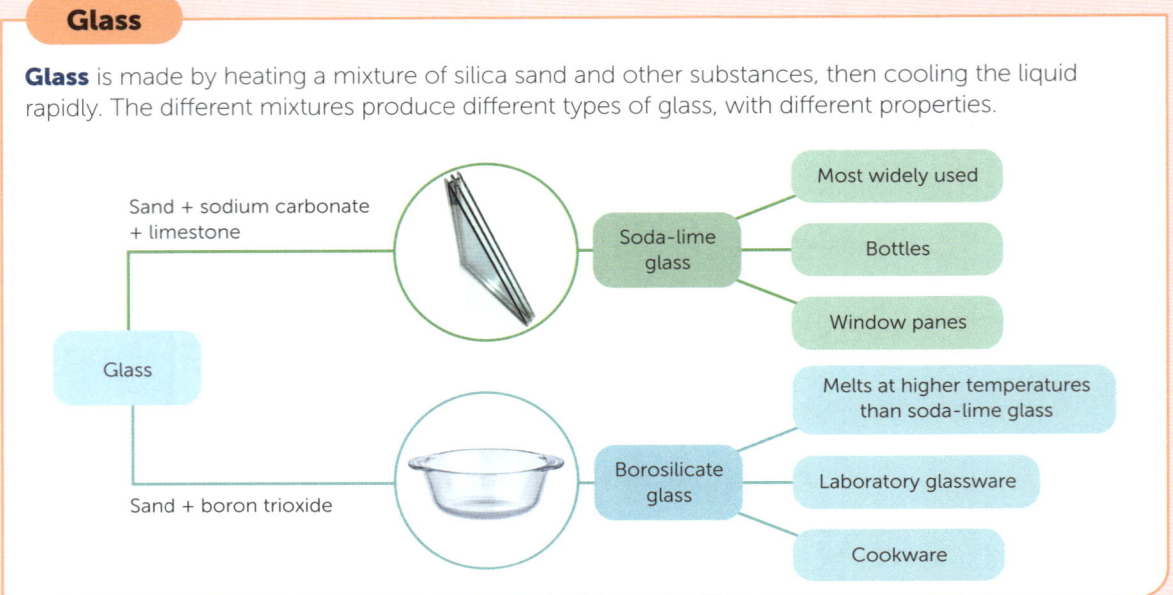

Clay ceramics

Pottery and bricks are examples of **clay ceramics.** They are made by shaping wet clay, then heating the clay to high temperatures in an oven.

Clay ceramics are heat-resistant, so they are used to make plates, laboratory crucibles and linings for the inside of kilns.

Recycling one tonne of glass releases around 250 kg less carbon dioxide than making the same mass of new glass.

---

1. Compare the properties of glass with those of clay ceramics. [4]
2. Suggest why soda-lime glass is more easily recycled than borosilicate glass. [2]

   1. Both materials are hard[1] but brittle (they break or shatter when hit)[1]. Glass is transparent but clay ceramics are usually opaque (you cannot see through them).[1] Both materials are dense, so items made from them are heavy for their size.[1]
   2. Soda-lime glass melts at a lower temperature than borosilicate glass[1], so it is more easily melted down and moulded into new objects[1].

4.10.3.3 Chemistry

COMPOSITES

Composites are made from different materials with contrasting properties.

Matrix and reinforcement

Composites are usually made from two materials: the matrix (also called the binder) and the reinforcement. The reinforcement can be fragments or fibres.

Composite material	Matrix	Reinforcement
Fibreglass	Polymer resin	Layers of matting made from thin glass fibres
Plywood	Wood glue	Sheets of wood
Concrete	Cement	Sand and small stones

The structure of concrete

Making fibreglass

Concrete itself may be further reinforced by embedding a mesh of steel rods in it. Concrete breaks when you try to stretch or bend it. Steel rod does not easily break when you try to stretch or bend it. Both materials resist being squashed. The composite resists being squashed, stretched and bent.

> Medium-density fibreboard (MDF) is a composite widely used in furniture. It consists of a resin glue matrix with a reinforcement of wood chips.

---

1. Horsehair plaster was used in the past to cover walls. It consists of horsehair loosely mixed with plaster, which is smoothed onto the wall and allowed to set hard.
   (a) Identify the matrix and the reinforcement in horsehair plaster. [2]
   (b) Suggest **one** reason why horsehair was added to the plaster instead of just using plaster. [1]
2. A computer keyboard consists of polymer keys embedded in a metal board.
   Explain why the keyboard is **not** a composite. [2]

   1. (a) The matrix is plaster[1] and the reinforcement is horsehair[1].
      (b) It makes a composite that is stronger / less likely to break than plaster alone.[1]
   2. The keys are free to be moved up and down separately inside the board[1] but the components of a composite must be stuck together[1].

AQA GCSE **Chemistry** 8462 / 8464 – Topic 10

4.10.4.1 Chemistry

THE HABER PROCESS 1

The **Haber process** is an industrial process that makes ammonia.

A reversible reaction

The Haber process involves a reversible reaction:

nitrogen + hydrogen ⇌ ammonia

This means that only some of the nitrogen and hydrogen will react to produce ammonia, and some of the ammonia will break down to form nitrogen and hydrogen again.

The Haber process was invented by a German chemist called Fritz Haber. Around 450 million tonnes of fertiliser are made each year using ammonia made by this process. You can revise fertilisers on **page 160**.

Reaction conditions

The conditions used in the Haber process are:
- A high temperature – about 450 °C
- A high pressure – about 200 atmospheres.

Iron is also used to act as a **catalyst** for the reaction. See catalysts on **page 95**.

Nitrogen and hydrogen are mixed together and passed into the reactor. Some of the mixture reacts to produce ammonia, and the remainder is recycled to give it another chance to react.

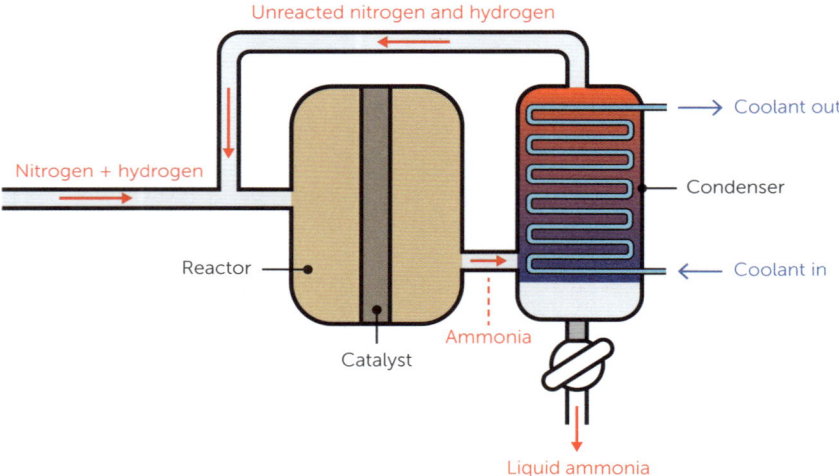

The ammonia is passed into a **condenser**, where it is cooled and condensed. The liquefied ammonia is removed and used to manufacture other substances such as fertilisers.

---

1. Name a source of nitrogen for the Haber process. [1]
2. Describe how hydrogen is produced as a raw material for the Haber process. [2]

   1. Air.[1]
   2. Coal or natural gas[1] is reacted with steam[1].

THE HABER PROCESS 2

4.10.4.1 | Chemistry | Higher Tier

Le Chatelier's principle can be applied to the Haber process.

Equilibrium

The relative amounts of all the reacting substances remain constant when a reversible reaction reaches equilibrium. Le Chatelier's principle can be used to predict the effect of changing the reaction conditions on the equilibrium position. See **pages 98–101**.

In the Haber process, the reacting substances are in the gas state:

$$N_2(g) + 3H_2(g) \rightleftharpoons 2NH_3(g)$$

Energy change of the forward reaction = −92 kJ/mol

The equilibrium **yield** of ammonia **increases** if:
- the temperature decreases
- the pressure increases
- ammonia is removed as it forms.

The iron catalyst does not change the equilibrium yield, but it does reduce the time taken to reach equilibrium.

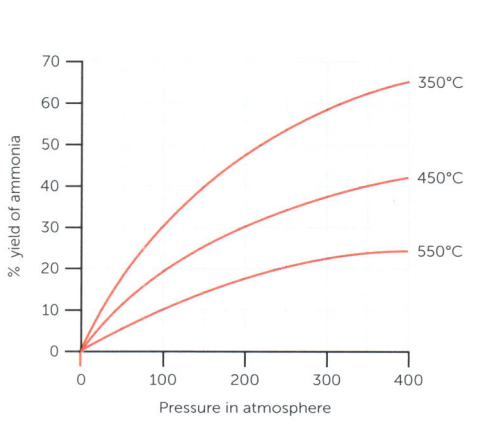

Steel becomes brittle when it is exposed to hydrogen at high pressures and temperatures. This makes it expensive to build reactors that need to withstand extreme conditions.

Reaction conditions

The temperature and pressure chosen for the Haber process are compromises:
- A lower temperature gives a higher equilibrium yield, but a higher temperature gives a greater rate of reaction.
- A higher pressure gives a higher equilibrium yield, but high pressures are expensive to maintain.

In addition to this, the reaction is not allowed to reach equilibrium. These conditions ensure a reasonable yield of ammonia in a reasonable time, at a commercially viable cost.

---

1. Explain why liquefying ammonia increases the equilibrium of ammonia. [3]
2. Refer to the graph in your answers to this question.
   (a) Determine the yield of ammonia at 450 °C and 200 atmospheres. [1]
   (b) The same yield could be achieved at 350 °C and 100 atmospheres.
      Explain why these conditions are not chosen. [2]

   1. Liquid ammonia leaves the mixture of reacting gases[1] which reduces the concentration of ammonia in the reaction mixture[1]. This shifts the equilibrium position to the right[1].
   2. (a) 30%[1]
      (b) The rate of reaction would be lower[1] so it may not be (as) profitable[1].

AQA GCSE **Chemistry** 8462 / 8464 – Topic 10

4.10.4.2 Chemistry

PRODUCTION AND USES OF NPK FERTILISERS

Fertilisers replace the mineral ions absorbed by plants as they grow.

NPK

The letters NPK in the term NPK fertiliser are the chemical symbols for nitrogen, potassium and phosphorus. These are three elements that plants need to grow well. A plant will suffer from deficiency diseases if it cannot obtain enough of these elements through its roots.

NPK fertilisers contain compounds of nitrogen, phosphorus and potassium. They prevent deficiency diseases and increase the yield of crops.

Cassava plants with potassium deficiency

Compounds in fertilisers

Plants can only absorb water and **soluble** substances through their roots. This means that nitrogen, potassium and phosphorus must be supplied as soluble salts.

Element	Ion(s) in salts	Typical salt in fertilisers
Nitrogen	nitrate, NO_3^- ammonium, NH_4^+	Ammonium nitrate
Phosphorus	phosphate, PO_4^{3-}	Ammonium phosphate
Potassium	potassium, K^+	Potassium sulfate

Notice that some salts contain more than one important ion.

Fertilisers are examples of **formulations**. Different salts are mixed in carefully measured proportions so that the fertiliser contains the appropriate percentage of the elements needed. You can revise formulations on **page 123**.

The different ions needed are obtained in different ways:
- Nitrate ions and ammonium ions from manufactured ammonia
- Phosphate ions by mining phosphate rock
- Potassium ions by mining rocks containing potassium compounds

Most salts containing phosphate ions are insoluble, but salts containing ammonium, nitrate or potassium ions are soluble in water.

---

1. (a) Name the industrial process which manufactures ammonia. [1]
   (b) Describe how the raw materials for making ammonia are obtained. [5]

   (a) Haber process[1].
   (b) The raw materials are nitrogen[1] and hydrogen[1]. Nitrogen is obtained from the air[1] and hydrogen is obtained by the reaction of coal or natural gas[1] with steam[1].

Making NPK fertilisers

NPK fertilisers are made from different raw materials using integrated processes.

Sources of nitrogen

Nitric acid, HNO_3, is made from ammonia. Overall:

$$\text{ammonia + oxygen + water} \rightarrow \text{nitric acid}$$

Ammonia and nitric acid react together to produce ammonium nitrate, NH_4NO_3.

Sources of potassium and phosphorus

Rocks containing potassium chloride or potassium sulfate may be used in fertilisers with little processing. Rocks containing phosphate compounds cannot be used without processing.

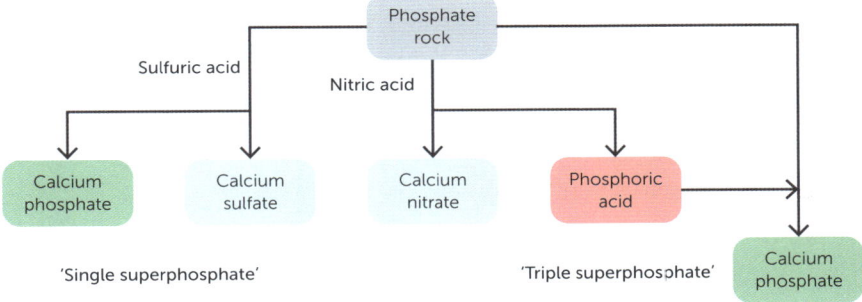

'Single superphosphate' and 'triple superphosphate' fertilisers both contain calcium phosphate, but triple superphosphate contains a much higher proportion of this phosphorus compound.

Production and processes

The industrial production of compounds used in fertilisers happens on a much greater scale than the laboratory preparation of the same compounds. It is a **continuous process** in which raw materials are supplied all the time, suitable reactions take place, and the products are separated and purified ready to make into fertilisers. These processes are rarely stopped.

A laboratory preparation is a batch process in which purified reactants are brought together so that they can react, then the products are separated and purified. Afterwards, the apparatus is cleaned ready for the next batch of compounds to be made.

---

2. Name the salts produced when phosphate rock is treated with:
   (a) sulfuric acid [2]
   (b) nitric acid [1]
   (c) phosphoric acid [1]

2. (a) Calcium phosphate[1] and calcium sulfate[1].
   (b) Calcium nitrate.[1]
   (c) Calcium phosphate.[1]

The raw materials for making sulfuric acid are sulfur, air and water. Sulfur is a useful by-product of removing it from hydrocarbon fuels to prevent sulfur dioxide pollution (you can revise atmospheric pollutants on **page 142**).

TOPIC 10

EXAMINATION PRACTICE

01 Fresh water is treated to make it potable or safe to drink.
    01.1 Give **one** reason why filtration is used to treat fresh water to make it potable. [1]
    01.2 Explain why chlorine may be added during water treatment. [2]
    01.3 Give **one** way in which sea water may be treated to make it potable. [1]
    01.4 Give a reason why sea water is not used to make potable water in the UK. [1]

02 The diagram shows three of the processes used to treat waste water.

Sedimentation → Anaerobic digestion → Aerobic biological digestion

    02.1 Describe what happens in the stage before sedimentation. [2]
    02.2 Describe the function of microbes in the treatment of waste water. [2]

03 Aluminium is obtained from aluminium ore or by recycling used aluminium items.
    03.1 Explain why aluminium ore is described as a limited resource. [2]
    03.2 Suggest **two** reasons, other than conserving supplies of aluminium ore, why we should recycle aluminium instead of making aluminium from bauxite. [2]

04 This question is about sustainable development.
    04.1 Describe what sustainable development means. [2]
    04.2 Shirts can be made from polyester or cotton fibres. Crude oil is the raw material for polyester. Suggest why using polyester alone is less sustainable than using a mixture of both types of fibre. [2]

05 Shopping bags can be made from paper or plastic. The table shows information from life cycle assessments of these bags.

	Paper bags	Plastic bags
Mass per bag in g	50	30
Source of raw material	Trees	Crude oil
Energy used to make one bag in kJ	135	120
Biodegradable	Yes	No
Recyclable	Yes	Yes
Reusable	No	Yes

Evaluate the use of paper to make bags compared to the use of plastic. Use information in the table and your own knowledge and understanding. [6]

Gold is extracted from gold ores taken from mines, and by phytomining.
    Describe how phytomining is carried out. [3]
    Suggest advantages of phytomining gold compared to extracting gold from ores. [2]

Chemistry only:

07 To reduce the chance of rusting, a steel body panel of a car is coated in zinc, then painted. The zinc and paint layers may be damaged when stones hit them during driving.
    07.1 Explain why the panel does not rust if the paint is chipped off. [2]
    07.2 Explain why the panel may not rust, even if the zinc coating is chipped off. [2]

08 Low carbon steel and high carbon steel are alloys.
    08.1 Compare the properties of low carbon steel and high carbon steel. [2]
    08.2 Give **one** important chemical property of steel that contains chromium and nickel. [1]

09 Performance car parts can contain a variety of polymers and carbon fibre.
    09.1 Compare the effects of heating on thermosoftening and thermosetting polymers. [1]
    09.2 Explain why carbon fibre is described as a composite. [2]

10 Design an experiment to show that both air and water are needed for rusting to happen. Include essential apparatus and how you would use the results. [6]

11 This question is about NPK fertilisers.
    11.1 Phosphorus is one of the three essential elements in NPK fertilisers. Name the other **two** essential elements in NPK fertilisers. [2]
    11.2 Compare the sources of raw materials for ammonia with the sources raw materials for phosphorus in fertilisers. [4]
    11.3 Name the salt produced when phosphate rock is treated with nitric acid. [1]

12 The Haber process manufactures ammonia from nitrogen and hydrogen.
    12.1 Give the temperature and pressure used in the Haber process. [2]
    12.2 Describe how ammonia is removed from the reaction mixture. [3]

Higher Tier only:

13 The graph shows how temperature and pressure affect the equilibrium yield of ammonia.

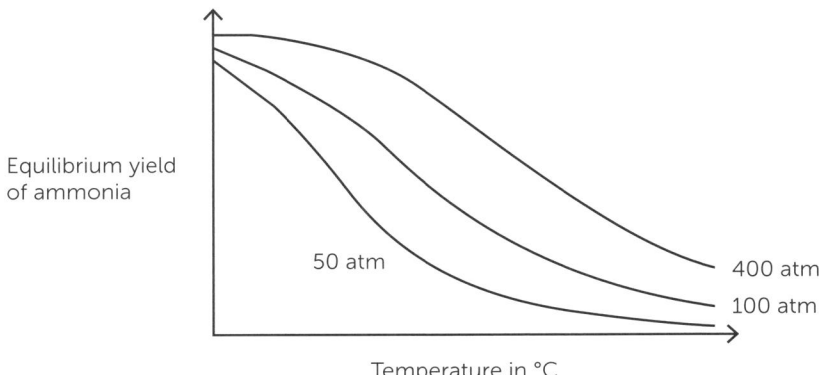

    13.1 Describe how the equilibrium yield of ammonia is affected by pressure.
    13.2 Evaluate the use of very high temperatures in the Haber process.
    13.3 Predict the effect on the graph of removing the iron catalyst from the reaction mixture. Explain your answer.

EXAMINATION PRACTICE ANSWERS

Topic 1

01 Any two from: [2]
- Compounds formed in chemical reactions, but mixtures not formed in chemical reactions.
- Compounds contain two or more elements chemically combined together, but substances in a mixture are not chemically combined together.
- Elements in compounds are present in fixed proportions, but substances in mixtures are in variable proportions.
- Chemical properties of elements in a compound are changed, but chemical properties of substances in mixtures are unchanged.
- Compounds are only separated into their elements by chemical reactions, mixtures can be separated by physical methods.

02 $2Fe_2O_3 + 3C \rightarrow 4Fe + 3CO_2$ Correctly balanced. [1]

03.1 It showed that atoms could be divided [1] leading to the plum pudding model [1] in which atoms are positive balls with negative electrons inside [1]. [2]

03.2 1 mark for each correct row to 3 marks: [3]

Name of subatomic particle	Relative charge	Relative mass
Neutron	0	1
Electron	−1	very small
Proton	+1	1

03.3 7 protons [1] 8 neutrons [1] 10 electrons [1]. [3]
03.4 2.8×10^{-10} m. [1]
04.1 relative atomic mass = $\dfrac{(69 \times 63) + (31 \times 65)}{(69 + 31)}$ [1] = $\dfrac{4347 + 2015}{100}$ [1] = 63.6 [1] 3 marks for correct answer without working. [3]
04.2 The chemical properties of an element are determined by the number of electrons [1]; the two isotopes have the same number of electrons / 29 electrons [1]. [2]
05.1 He left gaps for undiscovered elements [1]; he changed the order of some elements [1]. [2]
05.2 Group 5 [1] period 3 [1] (Allow top right for 1 mark only.) [2]
05.3 (Positively charged ions because) metals / transition metals are placed between groups 2 and 3 [1] and metals are elements that react to form positive ions / non-metal elements do not form positive ions [1]. [2]

06 Indicative content: [6]

Filtration
- to separate carbon particles from the ink
- as a residue

Fractional distillation
- of the filtrate
- to separate the propanol from the water

Chromatography
- of the filtrate / ink
- to separate the coloured substances from one another

Safety precautions relevant to the experiment with reasons
- eye protection because of solvent / hot liquid
- care with hot apparatus to avoid burns
- heat flask with an electrical heater / care with naked flames because ethanol is flammable

atoms of Group 0 elements have full outer shells / stable arrangements of electrons [1], so they have little tendency to electrons [1]. [2]
ng point increases going down the group / increases as the relative atomic mass increases. [1]
the range 3.65–3.90 g/m³ [1] because the relative atomic mass of krypton is about half-way between those xenon [1] and its density should be about half-way between their densities [1]. [3]
dicted to be a solid / a metal rather than a non-metal. [1]
NaOH(aq) + H_2(g) 1 mark for correct formulae, 1 mark for correct balancing, 1 mark for state symbols. [3]
tassium is further from the nucleus [1] so the potassium nucleus has a weaker force of attraction and the outer electron is lost more easily [1]. [3]

09.1 A halogen cannot displace itself from its salts. [1]

09.2 The order of decreasing reactivity is chlorine, bromine, iodine [1] because chlorine displaces bromine and iodine from their salts [1] and bromine displaces iodine but not chlorine [1]. [3]

09.3 $Br_2 + 2I^- \rightarrow 2Br^- + I_2$ [1] for correct formula, [1] for correctly balanced. [2]

10 **Chemistry only:** Indicative content: [6]

Evidence that zinc is not a transition metal
- transition metals form coloured compounds
- Group 1 metals form white compounds
- zinc hydroxide and zinc chloride are white not coloured
- zinc only forms Zn^{2+} ions
- transition metals form ions with different charges
- copper can form ions with different charges

Evidence that zinc is a transition metal
- zinc has a high melting point
- transition metals typically have high melting points
- zinc has a high density
- transition metals typically have high densities
- zinc oxide can act as a catalyst
- many transition metals act as catalysts

Topic 2

01.1 MgS. [1]

01.2 Electrons are transferred from the outer shell of a magnesium atom to the outer shell of a sulfur atom [1] forming Mg^{2+} ions [1] and S^{2-} ions [1]. [3]

01.3 (Strong) electrostatic forces of attraction [1] between oppositely charged ions [1]. [2]

01.4 Two from: high melting point [1], high boiling point [1], does not conduct electricity when solid [1], conducts electricity when molten/dissolved [1]. [2]

02.1 4 shared pairs of electrons [1] no other dots or crosses shown [1]. [2]

02.2 Methane exists as small molecules [1] which have weak forces between them [1] and which need relatively little energy to overcome [1]. [3]

03.1 Both contain carbon atoms joined by covalent bonds [1] and both forms giant structures [1]. Each carbon atom forms four covalent bonds in diamond but only three covalent bonds in graphite [1]. Graphite forms layers (of hexagonal rings of atoms) but diamond does not [1]. [4]

03.2 They have delocalised electrons [1] which can move and carry charge through the structures [1]. [2]

Answer for 02.1

04.1 Atoms in metals are held together by strong (metallic) bonding [1] which takes a lot of energy to overcome [1]. [2]

04.2 Its atoms are in layers [1] that can move over one another [1]. [2]

04.3 The lithium atoms distort the layered structure of the metal [1] which makes it more difficult for layers of atoms to move over one another [1]. [2]

04.4 Energy is transferred [1] by delocalised electrons [1]. [2]

05.1 Simplest whole number ratio [1] of atoms of each element in a substance [1]. [2]

05.2 FeO. [1]

06.1 (Addition) polymer. [1]

06.2 A covalent bond. [1]

06.3 They have relatively strong intermolecular forces [1] which need a lot of energy to overcome [1]. [2]

07 Indicative content: [6]

Section X	Section Y	Section Z
• dodecanoic acid is in solid state	• dodecanoic acid changes state from solid to liquid / melts	• dodecanoic acid is in liquid state
• particles in regular/lattice arrangement	• particles gain energy	• transition metals typically have high melting points
• vibrate about fixed positions	• energy used to overcome some of the molecular forces	• particles gain energy
• particles gain energy	• regular/lattice structure broken	• particles move around each other
• vibrate more rapidly	• arrangement of particles becomes random	• particles move randomly
		• movement increases

Chemistry only:

08.1 A particle 1–100 nm in size [1] consisting of a few hundred atoms [1]. [2]

08.2 Large surface area to volume ratio. [1]

08.3 Reduce the size of the nanoparticles by a factor of 10 [1] (but reduce size to side length 2.5 nm scores 2 marks). [2]

Topic 3

01.1 $(3 \times 24) + (2 \times 14) = 72 + 28 = 100$ (1 mark for correct answer without working). [1]
01.2 $56 + (2 \times 14) + (2 \times 3 \times 16) = 56 + 28 + 96 = 180$ (1 mark for correct answer without working). [1]
02   77.4% [2] Evidence of working [1] e.g. $\frac{3 \times 16}{62} \times 100$ [2]
03   (0.6 g of) carbon dioxide escaped from the boiling tube. [1]
04   Volume = $\frac{50}{1000}$ = 0.05 dm³ [1] Mass = 8.0 × 0.05 = 0.4 g [1] Correct answer scores 2 marks. [2]

Higher Tier only:

05   44 × 0.25 = 11 g. [1]
06   M_r of NaCl = (23 + 35.5) = 58.5 and M_r of Cl_2 = (2 × 35.5) = 71 [1].
      Amount of NaCl = $\frac{7.25}{58.5}$ = 0.124 mol [1]
      From the balanced equation, mole ratio NaCl : Cl_2 = 2 : 1 so 0.062 mol of Cl_2 [1]
      Mass of Cl_2 = 71 × 0.062 = 4.40 g [1] [4]
07   M_r of Fe_2O_3 = (2 × 56) + (3 × 16) = 160 [1].
      Amount of Mg = $\frac{960}{24}$ = 40 mol [1] Amount of Fe_2O_3 = $\frac{2.0 \times 1000}{160}$ = 12.5 mol [1]
      Amount of Mg needed = 3 × 12.5 = 37.5 mol which is less than the 40 mol added [1] [4]
08.1 Mass of O_2 = (6.85 – 6.69) = 0.16 g. [1]
08.2 Amount of PbO = $\frac{6.69}{223}$ = 0.03 mol. Amount of O_2= $\frac{0.16}{32}$ = 0.005 mol. Amount of Pb_3O_4= $\frac{6.85}{685}$ =0.01 mol [1]
      Simplest whole number ratio is 6 : 1 : 2 [1]
      $6PbO + O_2 \rightarrow 2Pb_3O_4$ [1] [3]

Chemistry only:

09   0.42 kg × 1000 = 420 g [1]
      Percentage yield = $\frac{420}{700} \times 100$ = 60% [1] (allow the alternative method in which 700 g is converted to 0.7 kg). [2]
10.1 Total relative formula mass of reactants = (2 × 18) = 36 [1]
      Percentage atom economy = $\frac{2 \times 2}{36} \times 100$ [1] = 11.1% [1] [3]
10.2 There are more products in Process 2. [1]
10.3 Sell one or both of the other products. [1]

Higher Tier only:

11   Volume = $\frac{250}{1000}$ = 0.25 dm³ [1] Amount of NaOH= $\frac{2.5}{40}$ = 0.0625 mol [1] Concentration= $\frac{0.065}{0.25}$ = 0.25 mol/dm³ [1] [3]
12   Amount of lead(II) nitrate = $\frac{25}{1000} \times 0.01$= 2.5 × 10⁻⁴ mol [1]
      From the balanced equation, mole ratio $PB(NO_3)_2$: KI is 1 : 2, so 2 × 2.5 × 10⁻⁴ = 5 × 10⁻⁴ mol KI [1]
      Concentration of KI = $\frac{1000}{10} \times 5 \times 10^{-4}$= 0.05 mol dm⁻³ [1] [3]
13   From the balanced equation, mole ratio O_2 : SO_3 is 1 : 2, so volume = 2 × 50 = 100 cm³. [1]
14.1 Volume of carbon dioxide= $\frac{84}{1000}$ =0.084 dm³ [1] Amount of carbon dioxide= $\frac{0.084}{24}$ [1] = 3.5 × 10⁻³ mol [1] [3]
14.2 Volume = $\frac{0.1}{44} \times 24$
      0.06 dm³ [1] (60 cm³) [1]

Topic 4

01.1 Z, X, Y. [1]
01.2 A metal cannot displace itself from its compounds. [1]
01.3 Magnesium was oxidised because it gained oxygen [1]. Copper(II) oxide was reduced because it lost oxygen [1]. [2]
02.1 Nitric acid. [1]
02.2 Copper does not react with dilute acids. [1]
02.3 Two from: copper(II) oxide [1], copper(II) hydroxide [1], copper(II) carbonate [1]. [2]

03.1 The pH of hydrochloric acid was less than 7 but the pH of the sodium hydroxide was more than 7. [1]
03.2 $H^+(aq) + OH^-(aq) \rightarrow H_2O(l)$ 1 mark for correct reactants correctly balanced, 1 mark for state symbols. [2]
04 Aluminium sulfide: Al_2S_3 [1], sodium phosphate Na_3PO_4 [1]. [2]
05.1 Aluminium is more reactive than carbon. [1]
05.2 Aluminium oxide is insoluble in water so it must be melted [1]. The molten mixture has a much lower melting point than aluminium oxide alone [1] which reduces the amount of energy needed for electrolysis to happen [1]. [3]
05.3 The positive electrode is made from carbon [1]. Oxygen is given off at the positive electrode [1] which reacts with the carbon (to form carbon dioxide) and wears it away [1]. [3]
06.1 It contains zinc ions and chloride ions [1] which are free to move [1]. [2]
06.2 Cathode: zinc [1] Anode: chlorine [1] . [2]
06.3 Chlorine is still given off at the anode [1] but hydrogen is given off [1] (instead of zinc). [2]

Higher Tier only:

07.1 $Cu^{2+} + 2e^- \rightarrow Cu$ [1]
07.2 $2Cl^- \rightarrow Cl_2 + 2e^-$ [1]
08.1 (Magnesium) because it loses electrons [1] $Mg \rightarrow Mg^{2+} + 2e^-$ [1]. [2]
08.2 Reduction and oxidation reactions happen at the same time. [1]
09.1 It is only partially ionised/dissociated [1] in aqueous solution [1]. [2]
09.2 pH 5. [1]

Chemistry only:

10.1 Mean titre = $\frac{(24.15 + 24.25)}{2}$ [1] = 24.20 cm³ [1]. [2]
10.2 The colour changes gradually [1] so you cannot be sure of the end point [1]. [2]

Higher Tier only:

11 Amount of NaOH = $\frac{25.0}{1000} \times 0.200 = 0.005$ mol [1].
   From the balanced equation, mole ratio H_2SO_4 : NaOH = 1 : 2 so 0.0025 mol of H_2SO_4 [1].
   Concentration of $H_2SO_4 = \frac{0.00250}{14.20} \times 1000 = 0.176$ mol/dm³ [1]. [3]

Topic 5

01 A reaction which transfers energy to the surroundings [1] causing the temperature of the surroundings to increase [1]. [2]
02.1 The process is endothermic [1] because the temperature goes down [1]. [2]
02.2 Named application that needs low temperatures, e.g. sports injury pack, ice pack for food. [1]
03 Indicative content: [6]
  - X is the activation energy
  - X is the energy needed for the reaction to happen
  - X is the energy needed to break bonds in $CaCO_3$
  - X is an endothermic process
  - X is 2191 kJ/mol
  - Y is the overall energy change
  - Y is 179 kJ/mol
  - Y is positive, so the reaction is endothermic
  - more energy needed to break bonds in reactants than is released when bonds form in the products
  - Z is the energy released when bonds form in CaO and CO_2
  - Z is an exothermic process
  - Z = (2191 – 179) = 2012 kJ/mol

04 Energy in to break bonds: (H–H) + 151 [1] Energy out when bonds form: (2 × 298) = 596 [1].
   Overall energy change = (H–H) + 151 – 596 = –9 [1] (H–H) = –9 – 151 + 596 = 436 kJ/mol [1]. [4]
05.1 The reaction in the lithium-ion battery is reversible [1] but the reaction in the alkaline battery is not reversible [1]. [2]
05.2 One of the reactants is used up. [1]
05.3 Change the zinc for a more reactive metal / change the copper for a less reactive metal [1], because the potential difference depends on the difference in reactivity between the two metals [1]. [2]
06.1 It provides oxygen [1] needed to oxidise the fuel/methanol [1]. [2]
06.2 $CH_3OH + H_2O \rightarrow 6H^+$ [1] + $6e^-$ [1] + CO_2 [1]

Topic 6

01. Increase the surface area / crush the lump to make a powder. [1]
02.1 One from: use a gas syringe [1] / use an inverted measuring cylinder. [1] [1]
02.2 Mean rate = $\frac{48}{60}$ [1] =0.8 [1] cm^3/s.[1] [3]
02.3 Line B [1] because calcium was the limiting reactant / hydrochloric acid was in excess [1], 1.5 times more calcium was used so 1.5 times more hydrogen would be produced [1]. [3]
02.4 **Higher Tier only:** Evidence of drawing a tangent at 40 s on line A and reading off values [1]. Calculations, for example:
Rate of reaction = $\frac{(100 \text{ cm}^3 - 70 \text{ cm}^3)}{(55 \text{ s} - 15 \text{ s})} = \frac{30 \text{ cm}^3}{40 \text{ s}}$ [1] = 0.75 [1] (cm^3/s) [3]
03. The minimum amount of energy that particles need to react. [1]
04.1 The concentration of reactants decreases [1] so the frequency of collisions decreases [1]. [2]
04.2 Reactant particles gain energy / move faster [1], the frequency of collisions increases [1], a greater proportion of collisions are successful / a greater proportion of reactant particles have the activation energy [1]. [3]
05.1 Vanadium(V) oxide [1] because it is present in the reaction mixture but does not appear in the balanced equation [1]. [2]
05.2 Catalysts increase the rate of reaction [1] by providing a reaction pathway with a lower activation energy [1]. [2]
06.1 It shows that the reaction is reversible. [1]
06.2 The colour would change from blue to pink. [1]
06.3 Heat it up. [1]
06.4 The reverse reaction is endothermic [1] and transfers the same amount of energy as the forward reaction [1]. [2]
07.1 The reacting substances must not escape for the reaction to reach equilibrium [1] but reactants/products can escape from an open boiling test tube [1]. [2]
07.2 It will be 0.50 g/s (the same) [1] because at equilibrium the rate of the forward reaction is exactly the same as the rate of the reverse reaction [1]. [2]
8. Indicative content: [6]

Variables to control
- mass of magnesium ribbon
- volume of hydrochloric acid
- surface area of magnesium ribbon
- temperature of reaction mixture

Method
- connect a bung and delivery tube to a gas syringe / upturned measuring cylinder in a trough of water
- use measuring cylinder to place a known volume of hydrochloric acid in a conical flask
- add a known mass of magnesium ribbon
- immediately start stopwatch and connect the flask to the delivery tube
- record the volume of gas at regular intervals

Analysis
- calculate the mean rates of reaction at each concentration of hydrochloric acid
- plot a graph of mean rate against concentration

Higher Tier only:

09. They increase the rate of reaction / reduce the time taken to reach equilibrium. [1]
10.1 (It increases the equilibrium yield) because there are fewer molecules of gas on the product side of the equation [1] so the equilibrium position moves to the right [1]. [2]
10.2 (It decreases the equilibrium yield) because increasing the temperature shifts the equilibrium position in the direction of the endothermic change [1] and this is away from dichloroethane [1]. [2]
11. The colour of the mixture will turn dark red [1] because the equilibrium position will move to the right [1] to form more products / away from the additional reacting substance [1]. [3]

Topic 7

01.1 It contains carbon and hydrogen only. [1]
01.2 Its name ends in ane [1] and its formula fits the general formula C$_n$H$_{2n+2}$ [1] [2]
02. 1 mark for correct number of each atom [1], 1 mark for all single bonds [1]: [2]
03.1 Two from: liquefied petroleum gases / LPG [1], petrol [1], diesel oil [1], kerosene [1], heavy fuel oil [1]. [2]

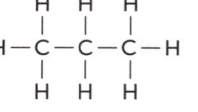

Answer for 02

03.2 Crude oil is heated to boil/evaporate it [1]; vapours pass up the fractionating column [1], they cool and condense [1] at different heights/temperatures [1]. [4]

04.1 It is a measure of how easily a substance flows. [1]

04.2 (As the molecules become larger) the boiling point increases [1] and the flammability decreases [1]. [2]

05 $C_5H_{12} + 8O_2 \rightarrow 5CO_2 + 6H_2O$. 1 mark for correct formulae, 1 mark for correct balancing. [2]

06 Mix with bromine water [1], bromine water changes from orange-brown to colourless with hexene but stays orange-brown / no change with hexane [1]. [2]

07.1 One from: to match supply of fractions with demand for them [1], to produce more fuels [1], to produce starting materials for polymers [1], to produce starting materials for other substances / petrochemicals [1]. [1]

07.2 Steam cracking needs higher temperatures [1] and higher pressures [1] than catalytic cracking. [2]

07.3 $\mathbf{2}C_8H_{18} \rightarrow C_7H_{16} + C_3H_8 + \mathbf{3}C_2H_4$ 1 mark for each balancing number. [2]

Chemistry only:

08.1 C_4H_8 [1]

08.2 Butene molecules contain two fewer hydrogen atoms than the alkane with four/same number of carbon atoms [1] (allow contain C=C bond for 1 mark) [1]

09
```
      H   H   H
      |   |   |
  H — C — C = C
      |       |
      H       H
```
[1]

10 They undergo incomplete combustion. [1]

11 Reaction with hydrogen happens at 150 °C / high temperature [1] with a nickel catalyst [1];
reaction with chlorine happens at room temperature [1]. [3]

12.1
```
      H   H
      |   |
  H — C — C — O — H
      |   |
      H   H
```
[1]

12.2 Two from: Yeast [1], atmospheric pressure [1], room temperature / stated temperature in the range 20–35 °C [1]. [2]

12.3 Bubbling [1] sodium disappears / colourless solution formed [1]. [2]

13.1 Its name ends in anoic acid / oic acid [1] its formula contains COOH [1]. [2]

13.2 Bubbling. [1]

13.3 Ester. [1] (Ethyl propanoate.) [1]

13.4 **Higher Tier only:** It forms solutions that have pH values less than 7 [1], it is only partially ionised/dissociated in solution [1]. [2]

14.1 C–C bond with no extra bonds [1], remainder of molecule correct [1], n added to the far left [1] [3]

14.2 It is formed from an alkene / monomer with a C=C bond. [1]

15.1 Two polymer chains [1], double helix [1]. [2]

15.2 DNA: (four different) nucleotides [1], starch: glucose / simple sugars [1]. [2]

Higher Tier only:

16.1 They produce a small molecule (as well as the polymer) [1] the small molecule is water [1]. [2]

16.2 Monomers for condensation polymers have two different functional groups / –OH and –COOH groups [1], monomers for addition polymers have C=C bonds [1]. [2]

17.1 Amino acid. [1]

17.2 $-\!\!\left(\text{HNCH(CH}_3\text{)CO}\right)\!\!-$ [1]

Topic 8

01 A **glowing** splint [1] held in the gas relights if the gas is oxygen [1]. [2]

02 Shake the gas with limewater / bubble the gas through limewater [1]. The limewater turns cloudy white / milky [1]. [2]

03 Collect a sample of each gas in a test tube [1]. (For the gas from the cathode) hold a burning splint near the test tube [1]. It ignites the gas with a pop sound if the gas is hydrogen [1]. (For the gas from the anode) hold **damp** litmus paper in the gas [1]. It turns white if the gas is chlorine [1]. [4]

04.1 In chemistry a pure substance only contains one element or compound / is not mixed with any other substance [1] In everyday use, a pure substance has nothing added to it / is in its natural state [1]. [2]

04.2 The components of an alloy are in measured quantities. [1]

05 $R_f = \dfrac{74 \text{ mm}}{120 \text{ mm}}$ [1] = 0.6166 [1] = 0.62 to 2 significant figures [1]. [3]

06 Measure its melting point [1], if the dodecanoic acid is pure it should melt at 43.8 °C / it will not melt at 43.8 °C if it is not pure [1]. [2]

07 Fertilisers are a mixture of components [1] in carefully measured quantities [1] each with a purpose / to make sure that the fertiliser has the required properties [1]. [3]

08 There is a mobile phase [1] and a stationary phase [1]; different substances distribute between the two phases by different amounts [1] so they move up the paper at different rates [1]. [4]

09 Indicative content: [6]

Apparatus
- chromatography paper
- beaker or other suitable container
- pencil and ruler
- propanol

Method
- draw a pencil line near the bottom of the paper
- write a spot of pen ink on the line
- place the paper in the solvent / propanone
- make sure the solvent is below the pencil line
- allow the solvent to move through the paper
- remove and dry the paper before the solvent reaches the top

Analysis
- examine the results
- ink containing one coloured substance produces one spot
- ink containing more than one coloured substance produces more than one spot

10.1 Carbon dioxide. [1]

10.2 Potassium [1] carbonate [1]. [2]

11.1 Test 1: Ca^{2+} / calcium [1]; Test 2: I^- / iodide [1]; Test 3: Fe^{3+} / iron(III) [1]. [3]

11.2 Solution Y does not contain sulfate ions [1] because sulfate ions give a (white) precipitate with acidified barium chloride solution [1]. [2]

12.1 To react with carbonate ions which may be present / to prevent carbonate ions interfering with the test [1] because they give a (white) precipitate in this test [1]. [2]

12.2 Hydrochloric acid contains chloride ions [1] which give a false positive result in this test [1]. [2]

13 **Two** from: more accurate [1], more sensitive [1], quicker [1]. [2]

14.1 Sodium [1] and strontium [1]. [2]

14.2 Flame emission spectroscopy can analyse mixtures of ions [1] but some flame colours are masked in flame tests involving mixtures [1]; or flame emission spectroscopy can be used to determine the concentrations of metal ions [1] but flame tests can only show the presence of metal ions [1]. [2]

Topic 9

01 Oxides of nitrogen [1], Water vapour [1]. [2]

02.1 Both atmospheres contain carbon dioxide [1] and nitrogen [1]. [2]

02.2 One from: Venus has more carbon dioxide [1] less nitrogen [1] less oxygen. [1]

03 Nitrogen four-fifths / 80% / 78% [1]; oxygen one-fifth / 20% / 21% [1]. [2]

04 Volcanoes gave off water vapour [1] which cooled and condensed [1]. [2]

05 carbon dioxide + water → glucose + oxygen. Reactants correct [1], products correct [1]. [2]

06 Photosynthesis [1] reduced the proportion of carbon dioxide [1] and increased the proportion of oxygen [1]. [3]

07.1 One from: methane [1], water vapour [1]. [1]

07.2 One from: use of hydrocarbon/fossil fuels [1], cement manufacture [1]. [1]

07.3 Greenhouse gases in the atmosphere let radiation from the Sun reach the Earth's surface [1], absorb radiation emitted by the Earth's surface [1] let short wavelength through but absorb long wavelength radiation [1]. [3]

08 Two from: more heat waves / fewer cold periods [1], more droughts / more rainfall [1], rising sea levels / coastal flooding [1], Melting ice caps / melting glaciers [1], Changes in seasons / farming practices [1]. [2]

09.1 The total amount of carbon dioxide (and other greenhouse gases) given off [1] over the life cycle of a product / service / event [1]. [2]

09.2 One sensible way to reduce individual carbon footprint, e.g. walk rather than take the car [1], turn off the lights when leaving a room [1], turn down heating [1]. [1]

10 One from: carbon dioxide [1], carbon monoxide [1], carbon [1]. [1]

11 Carbon monoxide [1], carbon [1], water (vapour) [1]. [3]

12 It is odourless [1] and colourless [1]. [2]

13 Carbon / pollution particulates. [1]

14 Indicative content: [6]

Apparatus
- fuels may contain sulfur
- sulfur reacts with oxygen
- when the fuel is used
- balanced equation, e.g. $S + O_2 \rightarrow SO_2$

Method
- in hot engines/furnaces
- nitrogen and oxygen
- in the air
- react together
- balanced equation, e.g. $N_2 + 2O_2 \rightarrow 2NO_2$

Analysis
- sulfur dioxide and oxides cause acid rain
- effect of acid rain, e.g. damage to trees / rivers / aquatic life / buildings
- sulfur dioxide and oxides cause respiratory problems

Topic 10

01.1 It removes insoluble solids from the water. [1]
01.2 It kills microbes / sterilises the water [1] which prevents disease caused by drinking water [1]. [2]
01.3 One from: distillation / simple distillation [1], reverse osmosis [1]. [1]
01.4 One from: treating sea water uses large amounts of energy [1], the UK has a lot of rainfall / fresh water [1]. [1]
02.1 The waste water is screened / grit is removed [1]. [2]
02.2 One from: bacteria carry out anaerobic digestion [1] of sewage sludge [1]; bacteria carry out aerobic biological treatment [1] of effluent [1]. [2]
03.1 Aluminium ore is no longer being made / is made very slowly [1] so it will run out if we keep using it [1]. [2]
03.2 Two from: less energy is used [1], reduces pollution (e.g. from mining) [1], landfill sites for waste material are running out [1], recycled aluminium is cheaper [1], recycling is more sustainable [1]. [2]
04.1 Meeting the needs of people today [1] without compromising the ability of people in the future to meet their own needs [1]. [2]
04.2 Two from: Polyester comes from a limited resource / crude oil is a limited resource [1], polyester is not biodegradable [1], cotton is a renewable resource [1], mixing cotton into the shirt reduces the amount of polyester needed [1]. [2]

05 Indicative content: [6]

Raw materials
- wood is a renewable resource
- plastics are made from crude oil
- crude oil is a limited resource
- cutting down trees harms the environment
- fewer trees can absorb less carbon dioxide
- drilling for oil harms the environment

Manufacturing
- paper bags need more energy
- more energy may cause more pollution
- carbon dioxide may be a pollutant
- carbon dioxide is a greenhouse gas

Transport
- paper bags are heavier
- more fuel needed to transport paper bags
- larger lorries may be needed

Use
- paper bags may break more easily
- more paper bags may be needed
- paper bags cannot easily be reused

Disposal
- paper bags decompose
- plastic bags do not decompose
- both types of bag may be recycled to make other products

06.1 Plants absorb gold / gold compounds [1], plants are harvested [1], plants are burned to produce ash containing gold / gold compounds [1]. [3]

06.2 Two from: less damage to the environment / less waste produced [1], less expensive / less energy needed [1], can extract smaller amounts of gold / gold from less concentrated sources [1], burning plants release useful energy [1], plants absorb carbon dioxide from the atmosphere as they grow [1]. [2]

07.1 The zinc layer is still there [1] to stop air and water reaching the metal [1]. [2]

07.2 Zinc is more reactive than iron/steel [1] and provides sacrificial protection [1]. [2]

08.1 High carbon steel is stronger / harder [1] (accept reverse for low carbon steel), high carbon steel is brittle but low carbon steel is easily shaped [1]. [2]

08.2 It resists corrosion/rusting. [1]

09.1 Thermosoftening polymers melt but thermosetting polymers do not. [1]

09.2 It consists of a matrix / resin [1] and a reinforcement / carbon fibres [1]. [2]

10 Indicative content: [6]

Essential apparatus
- iron or steel objects, e.g. nails
- test tubes with bungs
- water
- kettle / Bunsen burner
- oil
- calcium chloride / desiccant

Method
- boil then cool some water
- put a nail inside each container
- place calcium chloride in one container
- completely cover a nail with boiled water
- add a layer of oil over the boiled water
- half cover a nail with unboiled water
- stopper the containers and leave for several days
- inspect the nails for signs of rusting

Expected results
- only the nail half-covered by water should show signs of rusting

Analysis
- the nail in calcium chloride is dry but exposed to air
- the nail in boiled water is wet but not exposed to air
- the half-covered nail is exposed to both air and water

11.1 Nitrogen [1], potassium [1]. [2]

11.2 Ammonia is manufactured [1] using nitrogen from the air [1] and hydrogen from steam/coal/natural gas [1] but phosphorus comes from mined rocks [1]. [4]

11.3 Calcium nitrate. [1]

12.1 About 450 °C [1], about 200 atmospheres [1]. [2]

12.2 The reaction mixture is cooled [1], the ammonia liquefies/condenses [1], the liquid ammonia leaves the unreacted gases (nitrogen and hydrogen) [1]. [3]

13.1 The equilibrium yield increases as the pressure increases. [1]

13.2 The rate of reaction increases as the temperature increases [1] but the equilibrium yield decreases as the temperature increases [1], so very high temperature is not suitable [1]. [3]

13.3 The graph will look the same [1] because catalysts do not alter the equilibrium position of a reaction [1]. [2]

INDEX

Symbols
–COO– group 116
–COOH group 115, 118
–OH group 113, 118

A
acid rain 142
acids 62, 64, 67, 72, 115
activation energy 82, 93, 94
addition polymerisation 117
addition reactions 111
aerobic biological treatment 148
alcohols 112, 113
alkali metals 19
alkalis 67
alkanes 104
alkenes 107, 108, 109, 110
alkoxides 113
alloys 37, 154
alpha particle experiment 10
aluminium manufacture 75
amino acids 119
ammonia 30
anaerobic digestion 148
anions 129, 130
anode 76
anomalous readings 47
aqueous solution 3
atmosphere 135
atom 11
   electronic structure 14
atom economy 54
atomic
   model 10
   number 11
   radius 11
   weights 16
atom utilisation 54
Avogadro constant 48

B
balancing
   equations 3, 50
   numbers 50
ball and stick diagram 28, 29
bases 64
batteries 85
binary ionic compound 74
bio
   fuel 114
   leaching 149
   mass 145
boiling point 33, 122
bond energies 83
bonds 17, 26, 82
   breaking 107
bromine water 108
brass 154
bronze 154
Buckminsterfullerene 39
bulk property 33
Bunsen burner 66
butanoic acid 115
by-product 54

C
C=C bond 109
calibration curve 132
calorimeter 81
carbohydrates 119
carbonate ions 129
carbon dioxide
   equivalent 141
   test for 126
carbon
   footprint 141
   nanotubes 39
carboxylic acids 115, 116
catalyst 23, 92, 94, 95, 98
cathode 76
cations 128, 130
cells
   rechargeable 85
chemical
   analysis 130
   bonds 26
   cells 84
   equation 3
   measurements 47
chlorine
   reactions with 19
   test for 126
chromatography 7, 124, 125
clay ceramics 156
climate change 140
closed system 46
collision theory 93
colorimeter 91
coloured compounds 23
combustion 106, 110
composites 157
compound 2, 122
concentration
   calculations 55
   effects of 94
   of a solution 52
concordant titres 71
condensation polymers 118
conducting electricity 34, 37
conduction 81
conservation of mass 44
corrosion 152
covalent bonds 20, 26, 29, 30, 107
cracking 107, 108
crude oil 104
crystallisation 6, 66

D

delocalised electrons 31, 37
desalination 146
diamond 38
diesel 105
diol 118
disappearing cross investigation 91
displacement reactions 21, 60
displayed structural formula 29
distillate 8
   distillation 8, 147
DNA 119
dot and cross diagrams 27, 28, 30
double helix 119
dual-coding iii
ductile 37

E

Earth 135
effluent 148
electrodes 73, 77, 84
electrolysis 4, 34, 73, 74, 76, 77, 153
electrolyte 77, 84
electronic structure 14
electrons 11, 60
electron shell model 10, 14
electroplating 153
electrostatic forces 26, 34
element 2, 122
empirical formula 28
endothermic change 100
endothermic reactions 80, 82
energy
   changes in reactions 80
   levels 14
environmental impact 150
equilibrium 97, 101, 159
esters 116
ethanoic acid 115
ethanol 114
ethyl ethanoate 116
exothermic change 100
exothermic reactions 80, 81, 82
extracting metals 75
extraction and the reactivity series 61

F

feedstock 105
fermentation 114
fertilisers 160
filtration 5
finite resources 104, 145
flame
   emission spectroscopy 131, 132
   test 127, 130
formula 2
formulation 123, 160
fossil fuels 137
fractional distillation 9, 105
fractionating column 9
fresh water 146
fuel cells 85
fuels 105
fullerenes 39

G

gas 32
   common 126
   tests 77
giant
   covalent structures 36, 38
   ionic lattice 28, 34
   structures 31
glass 156
global
   dimming 142
   warming 139
gold 154
gradient (of the tangent) 89
graphene 39
graphite 38
graphs 89
   interpreting 100
greenhouse gases 138, 141
Group
   0 elements 18
   1 elements 19
   7 elements 20
groups of elements 15

H

Haber process 98, 158
half equations 4, 74
halide
   ions 129
   salts 21
haloalkanes 111
halogens 20
hardness 37
heavy fuel oil 105
homologous series 104, 109
hydration 112
hydrocarbons 104, 106
hydrochloric acid 62, 64, 72
hydrogen
   fuel cells 85
   test for 126
hypothesis 91

I

incomplete combustion 110
indicator (pH) 67
insoluble reactant 65
instrumental methods of analysis 131
ion 12, 23, 60, 63
ionic
   bonding 26, 27
   compounds 20, 28, 34, 63, 73, 74
   equation 4
   equations 4
   structures 28
isotopes 12

J

James Chadwick 10
J.J. Thomson 10

K

kerosene 105
kinetic energy 92

L

lattice 34, 36
law of conservation of mass 44
leachate 149
Le Chatelier's principle 98, 159
life cycle assessment (LCA) 150
limewater 129
limitations of the particle model 33
limiting reactant 51
liquid 32
LPG (liquefied petroleum gases) 105

M

malleable 37
manufacturing aluminium 75
mass 44, 48
mass number 12
matrix 157
mean rate (of reaction) 88
measurements 47
melting point 33, 122
Mendeleev 16
metal 17
   carbonates 64
   hydroxides 128
   oxides 61
metallic
   bonding 31
   bonds 26
methane 139
methanoic acid 115
mixtures 5, 122
mobile phase 7, 124
molar gas volume 56
mole 48
molecules 29
molten ionic compounds 74
monomers 35, 117

N

nail polish remover 116
nano
   particles 40
   tubes 39
natural gas 104
naturally occurring polymers 119
neutralisation 67, 68
   of acids 64
neutrons 11
Niels Bohr 10
nitric acid 64
noble gases 18
non-enclosed system 46
non-metals 17
non-rechargeable cells 85
NPK fertilisers 160
nuclear model 10

O

open systems 46
orders of magnitude 40
organic compounds 110
oxidation 59
oxygen, test for 126

P

paper chromatography 7
particle model 32, 33
   limitations 33
particles 12
pear drop sweets 116
percentage composition by mass 45
percentage yield 53
periodic table 2, 15
   development of... 16
periods 15
petrochemicals 105
petrol 105
photosynthesis 136
pH scale 67
phytomining 149
plum pudding model 10
pollutants 141, 142
polyesters 118
poly(ethene) 29, 155
polymers 29, 35, 105, 107, 155
polypeptides 119
potable water 146
precipitate 91
precipitation reactions 46, 128
pressure 92
   changes 101
product 2
propanoic acid 115
proteins 119
protons 11
pure substances 122
purifying water 147

R

radiation (solar) 138
range 47
rate of reaction 88, 90, 92, 97, 99
reactant 2
reaction
   pathways 54
   profile 82, 95
reactivity 15, 21
   series 60
rechargeable cells 85
recycling 151
redox reactions 59, 62
reduction 59
   of oxides 61
reinforcement 157
relative
   atomic mass 13
   formula mass 44
   mass 11
renewable resources 145
repeating unit (polymerisation) 117
resolution 47
respiration 136
reuse 151
reverse osmosis 146
reversible reactions 72, 96
R_f values 124
rusting 152

S

sacrificial protection 153
salt 62, 64, 66
salty water 146
saturated solution 6
scientific diagram 77
sea water 146
sedimentary rocks 137
separating mixtures 5
separation methods 9
sewage 148
shared electrons 29
sharing electrons 26
shells 14
silica 36
small molecules 35
sodium chloride structure 28
solid 32
solubility 6
soluble substance 5
solute 52
solvent 5, 116
space-filling diagram 28
state change 33
states of matter 32
state symbols 3
stationary phase 7, 124
steels 154
strong acids 72
subatomic particles 11
sulfates 130
sulfuric acid 62, 64, 72
surface area to volume ratio 40
sustainable development 54, 145

T

tangents 89
temperature 92
temperature
  effects of 94
  gradient 9
theoretical yield 53
thermal
  decomposition 46
  energy 37
thermosetting polymers 155
thermosoftening polymers 155
titrations 69, 70
titre 69
transition metals 22
turbidity 91

U

uncertainty in readings 47
universal indicator 67

V

viscosity 106
volume 52
  of gases 56, 90

W

waste water treatment 148
weak acids 72
word equation 2

Y

yeast 114
yield 53

NOTES, DOODLES AND EXAM DATES

Doodles

Exam dates

Paper 1:

Paper 2:

LEVELS BASED MARK SCHEME FOR EXTENDED RESPONSE QUESTIONS

What are extended response questions?

Extended response questions are worth 4, 5 or 6 marks. These questions are likely to have command words such as 'compare', 'describe', 'design', 'explain' or 'evaluate'. You need to write in continuous **prose** when you answer one of these questions. This means you must write in full sentences (rather than in bullet points), organised into paragraphs as necessary.

You may need to bring together skills, knowledge and understanding from two or more areas of the specification. To gain full marks, your answer needs to be logically organised, with ideas linked to give a sustained line of reasoning.

Some extended response questions involve calculations. These need two or more steps that must be done in the right order. These questions will use the command words 'calculate' or 'determine'.

Marking

Written answers are marked using 'levels of response' mark schemes. Examiners look for relevant points (indicative content) and use a best fit approach. This is based on your answer's overall quality and its fit to descriptors for each level. Calculations are given marks for each step shown.

Example level descriptors

Level descriptors vary, depending on the question being asked. Level 3 is the highest level and Level 1 is the lowest level. No marks are awarded for an answer with no relevant content. The table gives examples of the typical features that examiners look for.

Level	Marks	Descriptors for a method	Descriptors for an evaluation
3	5–6	The method would lead to a valid outcome. All the key steps are given, and they are ordered in a logical way.	The answer is detailed and clear. It includes a range of relevant points that are linked logically. The answer uses relevant data that may be given in the question. A conclusion is made that matches the reasoning in the answer.
2	3–4	The method might not lead to a valid outcome. Most of the key steps are given, but the order is not completely logical.	The answer is mostly detailed but not always clear. It includes some relevant points with an attempt at linking them logically. Data may not be used fully. A conclusion is given that may not fully match the reasoning given.
1	1–2	The method would not lead to a valid outcome. Some key steps are given, but they are not linked in a clear way.	The answer gives separate, relevant points. Uses little or no data that may be given in the question. The points made may be unclear. If a conclusion is given, it may not match the reasoning given in the answer.

COMMAND WORDS

A **command word** in a question tells you what you are expected to do.

The structure of a question

You should see one command word per sentence, with the command word coming at the start. A command word might not be used, however, if a question is easier to follow without one. In these cases, you are likely to see:

- What …?
- Why …?
- How …?

Command word	What you need to do
Balance	Add correct balancing numbers to a chemical equation.
Calculate	Use the numbers given to work out an answer.
Choose	Select from a range of options.
Compare	Write about **all** the similarities and/or differences between things.
Complete	Complete sentences by adding your answers in the spaces provided.
Define	Give the meaning of something.
Describe	Recall a fact, event or process accurately.
Design	Describe how something will be done, such as a practical method.
Determine	Use the data or information given to you to obtain an answer.
Draw	Produce a diagram, or complete an existing diagram.
Estimate	Work out an approximate value.
Evaluate	Use your knowledge and understanding, and the information supplied, to consider evidence for and against something. You must include a reasoned judgement in your answer.
Explain	Give the reasons why something happens, or make something clear.
Give, name, write	Only write a short answer, commonly just a single word, phrase or sentence.
Identify	Name or point out something.
Justify	Support your answer using evidence from the information given to you.
Label	Add the correct words or names to a diagram.
Measure	Use a ruler or protractor to obtain information from a photo or diagram.
Plan	Write a method.
Plot	Mark data points on a graph.
Predict	Write a likely outcome of something.
Show	Give structured evidence to come to a conclusion.
Sketch	Make an approximate drawing, such as a graph without axis units.
Suggest	Apply your knowledge and understanding to a new situation.
Use	You **must** base your answer on information given to you, otherwise you will not get any marks for the question. You might also need to use your own knowledge and understanding.

KEY TERMS IN PRACTICAL WORK

Experimental design

Key term	Meaning
Evidence	Measurements or observations collected using a valid method
Fair test	When the dependent variable is only affected by the independent variable
Hypothesis	A suggested explanation for observations or facts
Prediction	A reasoned statement that suggests what will happen in the future
Valid	A valid method involves fair testing and is suitable for an investigation
Valid conclusion	A discussion of a valid experiment and what it shows

Variables

A variable is a characteristic that can be measured or observed.

Type of variable	Meaning
Categoric	It has names or labels rather than values
Continuous	It has values rather than names or labels
Control	It affects the dependent variable, so it must be kept the same or monitored
Dependent	It is measured or observed each time the independent variable is changed
Independent	It is deliberately changed by the investigator

Measurements and measuring

Key term	Meaning
Accurate	Close to the true value
Calibrated	A device is calibrated when its scale is checked against a known value
Data	Measurements or observations that have been gathered
Interval	The measured gap between readings
Precise	Very little spread about the mean value
Range	The values between the measured maximum and minimum values
Repeatable	When the same results are obtained using the same method and apparatus
Reproducible	Someone else gets the same results, or when different apparatus and methods are used
Resolution	The smallest change a measuring device can show
True value	The value you would get in an ideal measurement
Uncertainty	An interval in which the true value will be found

Errors

Type of error	Meaning
Anomalous value	Anomalous results lie outside the range explained by random errors
Random	Unpredictably different readings – their effects are reduced by repeats
Systematic	Readings that differ from true values by the same amount each time
Zero	A type of systematic error where a device does not read 0 when it should

USEFUL EQUATIONS

Mathematical skills account for at least 30% of the marks in the exams. Foundation Tier students may be given equations like the ones below. All students may be given an unfamiliar equation if a question needs one.

Shown in the specification

$$\text{mean rate of reaction} = \frac{\text{quantity of reactant used}}{\text{time taken}}$$

$$\text{mean rate of reaction} = \frac{\text{quantity of product formed}}{\text{time taken}}$$

$$R_f = \frac{\text{distance moved by substance}}{\text{distance moved by solvent}}$$

Chemistry only

$$\% \text{ yield} = \frac{\text{mass of product actually made}}{\text{maximum theoretical mass of product}} \times 100$$

$$\% \text{ atom economy} = \frac{\text{relative formula mass of desired product from equation}}{\text{sum of relative formula masses of all reactants from equation}} \times 100$$

Not shown in the specification

relative formula mass (M_r) = sum of relative atomic masses (A_r) of atoms shown in the formula

relative atomic mass (A_r) using abundances of two isotopes, A and B =

$$\frac{(\text{mass number} \times \text{percentage}) \text{ of A} + (\text{mass number} \times \text{percentage}) \text{ of B}}{100}$$

mass of solute (g) = concentration of solution (g/dm^3) × volume of solution (dm^3)

Higher Tier only

mass (g) = amount (mol) × M_r

$$\text{concentration of solution (mol/dm}^3\text{)} = \frac{\text{amount of solute (mol)}}{\text{volume of solution (dm}^3\text{)}}$$

volume of gas at room temperature and pressure (dm^3) = amount of gas (mol) × 24

overall energy change of a reaction =
(energy in to break bonds in reactants) − (energy out when bonds form in products)

$$\text{gradient of a graph} = \frac{\text{change in vertical axis}}{\text{change in horizontal axis}}$$

THE PERIODIC TABLE

1	2											3	4	5	6	7	0
																	4 **He** Helium 2
7 **Li** Lithium 3	9 **Be** Beryllium 4					1 **H** Hydrogen 1						11 **B** Boron 5	12 **C** Carbon 6	14 **N** Nitrogen 7	16 **O** Oxygen 8	19 **F** Fluorine 9	20 **Ne** Neon 10
23 **Na** Sodium 11	24 **Mg** Magnesium 12											27 **Al** Aluminium 13	28 **Si** Silicon 14	31 **P** Phosphorus 15	32 **S** Sulfur 16	35.5 **Cl** Chlorine 17	40 **Ar** Argon 18
39 **K** Potassium 19	40 **Ca** Calcium 20	45 **Sc** Scandium 21	48 **Ti** Titanium 22	51 **V** Vanadium 23	52 **Cr** Chromium 24	55 **Mn** Manganese 25	56 **Fe** Iron 26	59 **Co** Cobalt 27	59 **Ni** Nickel 28	63.5 **Cu** Copper 29	65 **Zn** Zinc 30	70 **Ga** Gallium 31	73 **Ge** Germanium 32	75 **As** Arsenic 33	79 **Se** Selenium 34	80 **Br** Bromine 35	84 **Kr** Krypton 36
85 **Rb** Rubidium 37	88 **Sr** Strontium 38	89 **Y** Yttrium 39	91 **Zr** Zirconium 40	93 **Nb** Niobium 41	96 **Mo** Molybdenum 42	[98] **Tc** Technetium 43	101 **Ru** Ruthenium 44	103 **Rh** Rhodium 45	106 **Pd** Palladium 46	108 **Ag** Silver 47	112 **Cd** Cadmium 48	115 **In** Indium 49	119 **Sn** Tin 50	122 **Sb** Antimony 51	128 **Te** Tellurium 52	127 **I** Iodine 53	131 **Xe** Xenon 54
133 **Cs** Caesium 55	137 **Ba** Barium 56	139 **La** Lanthanum 57	178 **Hf** Hafnium 72	181 **Ta** Tantalum 73	184 **W** Tungsten 74	186 **Re** Rhenium 75	190 **Os** Osmium 76	192 **Ir** Iridium 77	195 **Pt** Platinum 78	197 **Au** Gold 79	201 **Hg** Mercury 80	204 **Tl** Thallium 81	207 **Pb** Lead 82	209 **Bi** Bismuth 83	[209] **Po** Polonium 84	[210] **At** Astatine 85	[222] **Rn** Radon 86
[223] **Fr** Francium 87	[226] **Ra** Radium 88	[227] **Ac** Actinium 89	[261] **Rf** Rutherfordium 104	[262] **Db** Dubnium 105	[266] **Sg** Seaborgium 106	[264] **Bh** Bohrium 107	[277] **Hs** Hassium 108	[268] **Mt** Meitnerium 109	[271] **Ds** Darmstadtium 110	[272] **Rg** Roentgenium 111	[285] **Cn** Copernicium 112	[286] **Nh** Nihonium 113	[289] **Fl** Flerovium 114	[289] **Mc** Moscovium 115	[293] **Lv** Livermorium 116	[294] **Ts** Tennessine 117	[294] **Og** Oganesson 118

Notes

The Lanthanides (atomic numbers 59–71) and the Actinides (atomic numbers 90–103) are omitted.

Relative atomic masses for Cu and Cl have not been rounded to the nearest whole number.

Key

Relative atomic mass
Atomic symbol
Name
Atomic (proton) number

EXAMINATION TIPS

When you practise examination questions, work out your approximate grade using the following table. This table has been produced using a rounded average of past examination series for this GCSE. Be aware that boundaries vary by a few percentage points either side of those shown.

GCSE Chemistry

Grade	9	8	7	6	5	4	3	2	1	U
F Tier (%)					65	56	41	26	11	0
H Tier (%)	74	64	55	43	35	26	18			

Combined Science: Trilogy

Grade	5–5	5–4	4–4	4–3	3–3	3–2	2–2	2–1	1–1	U
F Tier (%)	59	54	50	44	37	31	25	19	13	0

Grade	9–9	9–8	8–8	8–7	7–7	7–6	6–6	6–5	5–5	5–4	4–4	4–3	3–3
H Tier (%)	66	62	58	53	49	44	40	35	31	26	22	19	14

1. Read questions carefully. This includes any information such as tables, diagrams and graphs.

2. Remember to cross out any work that you do not want to be marked.

3. Answer the question that is there, rather than the one you think should be there. In particular, make sure that your answer matches the command word in the question. For example, you need to recall something accurately in a describe question but not say why it happens. However, you do need to say why something happens in an explain question.

4. All the examination papers will include multiple-choice questions (MCQs). Make sure you tick the correct number of boxes, or link boxes with straight lines. When completing sentences, use words from the word list if one is given.

5. Show all the relevant working out in calculations. If you go wrong somewhere, you may still be awarded some marks if the working out is there. It is also much easier to check your answers if you can see your working out. Remember to give units when asked to do so.

6. Plot the points on graphs to within half a small square. Lines of best fit can be curved or straight, but must ignore anomalous points. If the command word is sketch rather than plot, you only need to draw an approximate graph, not an accurate one.

7. Make sure you do not mix words and symbols in chemical equations. You will be given full marks if you are asked to write a word equation, but you give the correct balanced equation instead. This does not work the other way round! Check that all the numbers of atoms, ions and charges balance in symbol equations. Remember to include state symbols when asked.

8. Remember that you may be asked to label a diagram or to complete a diagram. Sometimes you may be given the words to use. Make sure you can recall experiments you have done.

Good luck!

Revision, re-imagined
the Clear**Revise** family expands

New titles coming soon!

These guides are everything you need to ace your exams and beam with pride. Each topic is laid out in a beautifully illustrated format that is clear, approachable and as concise and simple as possible.

They have been expertly compiled and edited by subject specialists, highly experienced examiners, industry professional and a good dollop of scientific research into what makes revision most effective. Past examinations questions are essential to good preparation, improving understanding and confidence.

- Hundreds of marks worth of examination style questions
- Answers provided for all questions within the books
- Illustrated topics to improve memory and recall
- Specification references for every topic
- Examination tips and techniques
- Free Python solutions pack (CS Only)

Absolute clarity is the aim.

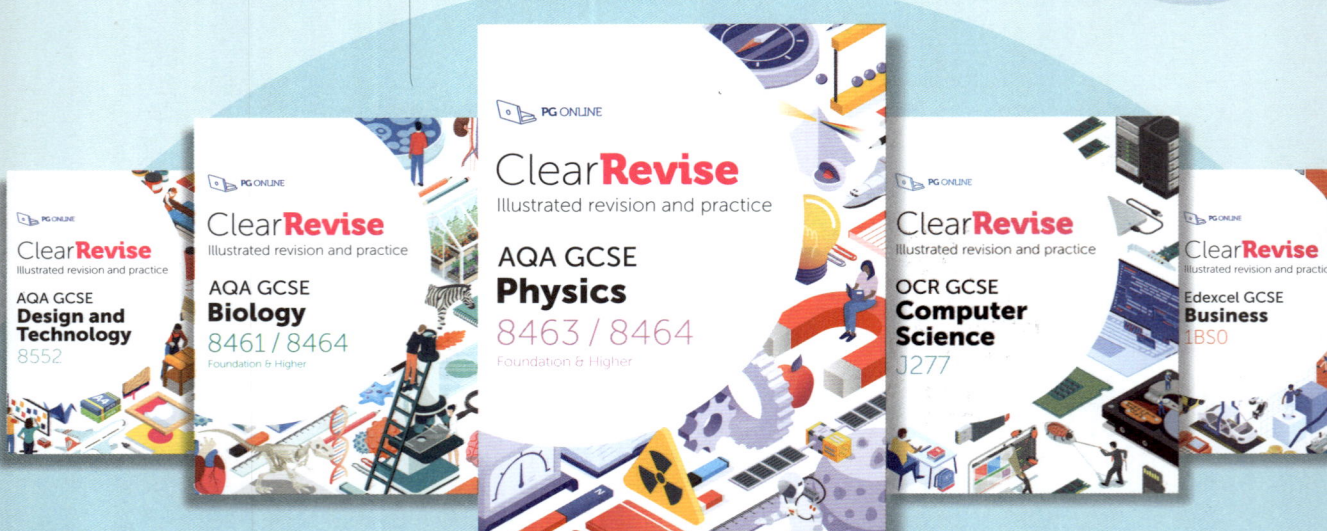

Explore the series and add to your collection at **www.clearrevise.com**

Available from all good book shops.

   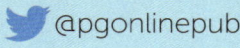 @pgonlinepub